Lecture Notes in Electrical Engineering

Volume 171

For further volumes:
http://www.springer.com/series/7818

Maxime Pelcat · Slaheddine Aridhi
Jonathan Piat · Jean-François Nezan

Physical Layer Multi-Core Prototyping

A Dataflow-Based Approach for LTE eNodeB

Maxime Pelcat
INSA Rennes
20 Avenue des Buttes de Coësmes,
CS 70839
35708 Rennes Cedex 7
France

Slaheddine Aridhi
Texas Instruments
Avenue Jack Kilby 821
06271 Villeneuve-Loubet
France

Jonathan Piat
LAA/CNRS
Avenue du colnel roche 7
31077 Toulouse
France

Jean-François Nezan
INSA Rennes
20 Avenue des Buttes de Coësmes,
CS 70839
35708 Rennes Cedex 7
France

ISSN 1876-1100 ISSN 1876-1119 (electronic)
ISBN 978-1-4471-6105-9 ISBN 978-1-4471-4210-2 (eBook)
DOI 10.1007/978-1-4471-4210-2
Springer London Heidelberg New York Dordrecht

Softcover reprint of the hardcover 1st edition 2013

Printed on acid-free paper

Springer is part of Springer Science+Business Media (www.springer.com)

Foreword

I am delighted to introduce this book, which covers important advancements to the state of the art in design methodologies for embedded signal processing systems. The targeted application area of long-term evolution (LTE) base stations is a major driver for modern wireless communication systems. By enabling more efficient embedded system implementations for LTE base stations, the book contributes powerful methods for lowering design, deployment, and maintenance costs for a wide range of wireless data networks.

The book's focus on multi-core signal processor technologies is also timely. The increasing data rates and quality-of-service requirements in wireless communications services place severe demands on base station processing, and require intensive parallel processing. To support such parallel signal processing implementations, multi-core digital signal processors (DSPs) have been evolving rapidly in recent years, with increasing amounts of hardware resources that can be operated simultaneously to provide enhanced tradeoff spaces between performance and power consumption.

To effectively utilize the parallelism provided by multi-core DSPs, designers of embedded signal processing systems face a challenging programming task, and the full potential of this technology is therefore difficult to approach on complex applications. To tackle this challenge, the authors introduce the Parallel and Realtime Embedded Executives Scheduling Method (PREESM). PREESM is a novel computer-aided design tool for parallel implementation of signal processing systems. PREESM applies innovative dataflow-based design methods for modeling signal processing applications, and couples these methods with features for rapid prototyping and code generation on multi-core DSP devices. Models and techniques for adaptive scheduling of dataflow graphs are also employed to provide robust execution of dataflow tasks on the targeted devices.

This book has evolved through a significant body of scientific contributions that the authors have published in technical papers and Ph.D. theses, with the core contributions evolving from the thesis of Dr. Maxime Pelcat. The work has been based on a rigorously integrated exploration of wireless communication system design, dataflow-based application design methodologies, embedded signal processing implementation, and advanced software tool development. The book provides a comprehensive view of the foundations of this exploration, as well as the novel design methods that have emerged from it, and demonstrations of their practical impact on multi-core system design for LTE. The book provides a major advance in design methods and tools for streamlined implementation of signal processing systems.

College Park, MD, USA, December 2011 Shuvra S. Bhattacharyya

Preface

The Internet communication service is being extended beyond its traditional frontiers of fixed wired infrastructure through the gradual addition of a broad range of complex networks and autonomous devices. In particular, the introduction of the Smartphone and subsequently the Tablets has produced a demand for mobile data services that has been growing rapidly. From traditional wireless networks to opportunistic networks of mobile devices in dense environments, new hardware platform requirements are becoming increasingly more complex, due to the necessary processing and computation power. The need for revolutionary or evolutionary architectures with multiple processors designed using different types of processors and technologies, and thus managing a range of heterogeneous components in order to support new multimedia services are only a subset of the wide range of existing new technologies that require high levels of processing power.

The consequence of the evolution of the mobile wireless standards is an increased need for the system to support multiple standards and multi-component devices for backward compatibility. These two requirements greatly complicate the development of telecommunication systems, imposing the optimization of device parameters over numerous constraints, such as performance, area and power. Achieving device optimization requires a deep understanding of application complexity and the choice of an appropriate architecture to support this application.

Of particular note, the new, feature-rich wireless standard called long-term evolution (LTE) is a complex application that needs a large amount of processing power. LTE is the next evolutionary step after 3G for mobile wireless communication, and is aimed at increasing the wireless network capacity while improving the spectral efficiency. LTE unites many technological innovations from diverse research areas such as digital signal processing, Internet protocols, network architecture, and security, and, as such, will drastically change the way that the worldwide mobile network is used in the future. LTE is anticipated to be the first truly global wireless standard, as it may be deployed in a variety of spectrum and operating scenarios, and has the capability to support a myriad of wireless applications. Numerous operators and service providers around the world have already deployed LTE on their networks or have announced LTE as their intended next generation technology.

An LTE eNodeB or base station must use powerful embedded hardware platforms, to offer a complete feature set with reasonable cost and power. Multi-core digital signal processors (DSP) combine cores with processing flexibility and hardware coprocessors that accelerate complex functionalities or repetitive processes, and, as such, are suitable hardware architectures to execute complex operations in real-time. The focus of this study is the physical layer portion of LTE eNodeB, so to understand the dynamism and the parallelism of this application under specific constraints such as latency.

The novel method of rapid prototyping was designed to alleviate the long simulation times now required in device optimization for complex architectures. It consists of studying the design tradeoffs at several stages of the development, including the early stages, when the majority of the hardware and software is not available. Initially, the inputs to a rapid prototyping process must be models of system parts, and are much simpler than in the final implementation. In a perfect design process, programmers would enlarge the models progressively, heading towards the final implementation.

Imperative languages, particularly C, are presently the preferred languages to program DSPs. Decades of compilation optimizations have refined these languages resulting in a good tradeoff between readability, optimality and modularity. However, imperative languages have been developed to address sequential hardware architectures inspired by the Turing machine and their ability to express algorithm parallelism is limited. Alternatively, dataflow languages and models have proven over many years to be efficient representations of parallel algorithms, thus allowing the simplification of their analysis.

This work proposes a novel top-down approach to tackle the multicore programming issue; it uses the rapid prototyping method and dataflow language for the simulation and code generation processes. Building on the base generated by other authors, the rapid prototyping method, which permits study of LTE signal processing algorithms, is constructed and explained. The relevant signal processing features of the physical layer of the 3GPP LTE telecommunication standard are then detailed. The next important building block in this novel approach is the dataflow model of computation, which is the model used to describe the application algorithms in this study. This new methodology is then contrasted with the existing techniques for system programming and rapid prototyping.

The System-Level Architecture Model (S-LAM) is a topology graph which defines data exchanges between the cores of a heterogeneous architecture. This S-LAM architecture model is developed to feed the rapid prototyping method. Additionally, a scheduler structure is detailed; this structure allows the separation of the different aspects of multi-core scheduling, in addition to providing other improvements to state-of-the-art methods. The scheduler developed in this study has the capacity to schedule the dynamic algorithms of LTE at run-time. The subsequent analysis using the methodology based on these blocks is two-fold:

firstly the LTE rapid prototyping and simulation, and then the code generation. The code generation is divided into two parts; the first deals with static code generation, and the second with the adaptive scheduler driven by a multi-core operating system.

The role of this work is to contribute a new methodology to a problem that will be ever more present. Software rapid prototyping methods were developed to replace certain tedious and suboptimal steps of the current test-and-refine methodologies for embedded software development. In this study, these new techniques were successfully applied to 3GPP LTE eNodeB physical layers. As such, this work has an important place in the study of optimal deployment of dynamic parallel applications, particularly those targeting LTE eNodeB lower layers, onto multi-core-based device architectures.

Nice, France, December 2011 Dr. Slaheddine Aridhi

Contents

Chapter 1
Introduction

1.1 Overview

The recent evolution of digital communication systems (voice, data and video) has been dramatic. Over the last two decades, low data-rate systems (such as dial-up modems, first and second generation cellular systems, 802.11 Wireless local area networks) have been replaced or augmented by systems capable of data rates of several Mbps, supporting multimedia applications (such as DSL, cable modems, 802.11b/a/g/n wireless local area networks, 3G, WiMax and ultra-wideband personal area networks). One of the latest developments in wireless telecommunications is the `3GPP Long Term Evolution` (LTE) standard. LTE enables data rates beyond hundreds of Mbit/s.

As communication systems have evolved, the resulting increase in data rates has necessitated higher system algorithmic complexity. A more complex system requires greater flexibility in order to function with different protocols in diverse environments. In 1965, Moore observed that the density of transistors (number of transistors per square inch) on an integrated circuit doubled every two years. This trend has remained unmodified since then. Until 2003, the processor clock rates followed approximately the same rule. Since 2003, manufacturers have stopped increasing the chip clock rates to limit the chip power dissipation. Increasing clock speed combined with additional on-chip cache memory and more complex instruction sets only provided increasingly faster single-core processors when both clock rate and power dissipation increases were acceptable. The only solution to continue increasing chip performance without increasing power consumption is now to use `multi-core chips`.

A `base station` is a terrestrial signal processing center that interfaces a radio access network with the cabled backbone. It is a computing system dedicated to the task of managing user communication. It constitutes a communication entity integrating power supply, interfaces, and so on. A base station is a real-time system because it treats continuous streams of data, the computation of which has hard time constraints. An LTE network uses advanced signal processing features including

M. Pelcat et al., *Physical Layer Multi-Core Prototyping*,
Lecture Notes in Electrical Engineering 171, DOI: 10.1007/978-1-4471-4210-2_1,

Orthogonal Frequency Division Multiplexing Access (OFDMA), Single Carrier Frequency Division Multiplexing Access (SC-FDMA), Multiple Input Multiple Output (MIMO). These features greatly increase the available data rates, cell sizes and reliability at a cost of an unprecedented level of processing power. An LTE base station must use powerful hardware platforms. Multi-core Digital Signal Processors (DSP) are especially designed to execute the complex signal processing operations in real-time optimizing the device power consumption. They combine cores with processing flexibility and hardware coprocessors that accelerate repetitive processes.

The consequence of evolution of the standards and parallel architectures is an increased need for the system to support multiple standards and multicomponent devices. These two requirements complicate much of the development of telecommunication systems, imposing the optimization of device parameters over varying constraints, such as performance, area and power. Achieving this system optimization requires a good understanding of the application complexity and the choice of an appropriate architecture to support this application. `Rapid prototyping` consists of studying the design tradeoffs at several stages of the development, including the early stages, when the majority of the hardware and software are not available. The inputs to a rapid prototyping process must then be models of system parts, and are much simpler than in the final implementation. In a perfect design process, programmers would refine the models progressively, heading towards the final implementation.

Imperative languages, and C in particular, are presently the preferred languages to program DSPs. Decades of compilation optimizations have made them a good tradeoff between readability, optimality and modularity. However, imperative languages have been developed to address sequential hardware architectures inspired on the Turing machine and their ability to express algorithm parallelism is limited. Over the years, `dataflow languages` and models have proven to be efficient representations of parallel algorithms, allowing the simplification of their analysis. A dataflow description is a directed graph where vertices (or actors) process data and edges carry data, with the requirement that vertices cannot share data. Actors can only communicate with other actors through ports connected to edges. In 1978, Ackerman explains the effectiveness of dataflow languages in parallel algorithm descriptions [1]. He emphasizes two important properties of dataflow languages:

- `data locality`: data buffering is kept as local and as reduced as possible,
- `scheduling constraints reduced to data dependencies`: the scheduler that organizes execution has minimal constraints.

The semantics of a dataflow program are defined by a Model of Computation (MoC) that dictates conditions for existence of a valid schedule, bounded memory consumption, proof of termination, and other properties. MoCs go from Synchronous Dataflow (SDF) with total compile-time predictability with respect to scheduling, memory consumption, termination, to dynamic dataflow where those properties are not predictable in the general case, with increasing expressiveness. The research on new MoCs and the use of heterogeneous mixtures of MoCs is still very active and the results presented in this book are part of this research fields.

The absence of remote data dependency simplifies algorithm analysis and helps to create a dataflow code that is correct-by-construction. The minimal scheduling constraints express the algorithm parallelism maximally. However, good practises in the manipulation of imperative languages to avoid recalculations often go against these two principles. For example, iterations in dataflow redefine the iterated data constantly to avoid sharing a state where imperative languages promote the shared use of registers. But these iterations conceal most of the parallelism in the algorithms that must now be exploited in multi-core DSPs. Parallelism is obtained when functions are clearly separated and Ackerman gives a solution to that: "to manipulate data structures in the same way scalars are manipulated". Instead of manipulating buffers and pointers, dataflow models manipulate `tokens`, abstract representations of a data quantum, regardless of its size.

It may be noted that digital signal processing consists of processing streams (or flows) of data. The most natural way to describe a signal processing algorithm is a graph with nodes representing data transformations and edges representing data flowing between the nodes. The extensive use of Matlab Simulink is evidence that a graphically editable plot is suitable input for a rapid prototyping tool.

The 3GPP LTE is the first application prototyped using the Parallel and Real-time Embedded Executives Scheduling Method (`PREESM`). PREESM is a rapid prototyping tool with code generation capabilities initiated in 2007 and developed with the first main objective of studying LTE physical layer. For the development of this tool, an extensive literature survey yielded much useful research: the work on dataflow process networks from University of California, Berkeley, University of Maryland and Leuven Catholic University, the Algorithm-Architecture Matching (AAM) methodology and SynDEx tool from INRIA Rocquencourt, the multi-core scheduling studies at Hong Kong University of Science and Technology, the dynamic multithreaded algorithms from Massachusetts Institute of Technology among others.

PREESM is a framework of plug-ins rather than a monolithic tool. PREESM is intended to prototype an efficient multi-core DSP development chain. One goal of this study is to use LTE as a complex and real use case for PREESM. In 2008, 68 % of DSPs shipped worldwide were intended for the wireless sector [2]. Thus, a multi-core development chain must efficiently address new wireless application types such as LTE. The term multi-core is used in the broad sense: a base station multi-core system can embed several interconnected processors of different types, themselves multi-core and heterogeneous. These multi-core systems are becoming more common: even mobile phones are now such distributed systems.

While targeting a classic single-core Von Neumann hardware architecture, it must be noted that all DSP development chains have similar features, as displayed in Fig. 1.1a. These features systematically include:

- A textual language (C, C++) compiler that generates a sequential assembly code for functions/methods at compile-time. In the DSP world, the generated assembly code is native, i.e. it is specific and optimized for the Instruction Set Architecture (ISA) of the target core.
- A linker that gathers assembly code at compile-time in an executable code.

Fig. 1.1 Comparing a present single-core development chain to a possible development chain for multi-core DSPs

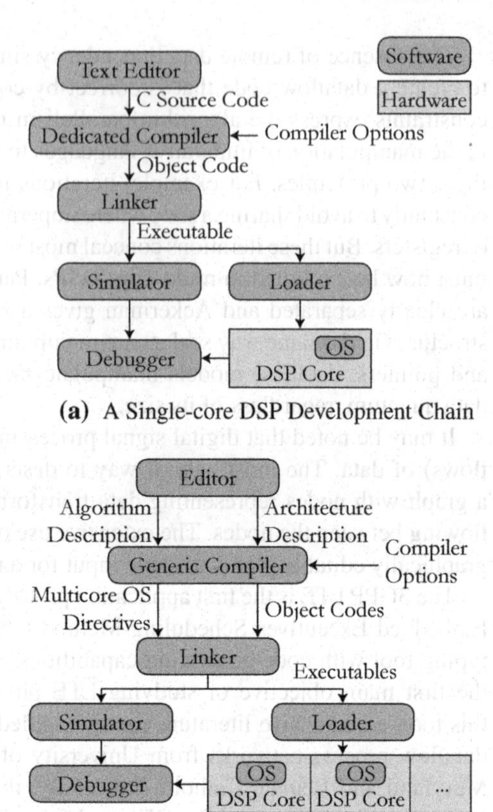

- A simulator/debugger enabling code inspection.
- A loader placing programs into memory and preparing them for execution.
- An Operating System (OS) that launches the processes, each of which comprise several threads. The OS handles the resource shared by the concurrent threads.

Incontrast with the DSP world, the generic computing world is currently experiencing an increasing use of bytecode. A bytecode is more generic than native code and is Just-In-Time (JIT) compiled or interpreted at run-time. It enables portability over several ISA and OS at the cost of lower execution speed. Examples of JIT compilers are the Java Virtual Machine (JVM) and the Low Level Virtual Machine (LLVM). Embedded systems are defined as computer systems designed for specific control functions within a larger system. Embedded systems are thus dedicated to a single functionality and in such systems, compiled code portability is not currently advantageous enough to justify performance loss. It is thus unlikely that JIT compilers will appear in DSP systems soon. However, as embedded system designers often have the choice between many hardware configurations, a multi-core development chain must have the capacity to target these hardware configurations at compile-time. As a consequence, a multi-core development chain needs a complete input architecture model instead of a few textual compiler options, such as used in single-core development chains. Extending the structure of Fig. 1.1a, a development chain for multicore DSPs may be imagined with an additional input architecture model (Fig. 1.1b). This multi-core development chain generates an executable for each core in addition to directives for a multi-core OS managing the different cores at run-time according to the algorithm behavior.

The present multi-core programming methods generally use test and refine methodologies. When processed by hand, parallelism extraction is hard and error-prone, but potentially extremely optimal (depending on the programmer). Programmers of Embedded multi-core software will only use a `multi-core development chain` if:

- The development chain is configurable and is compatible with the previous programming habits of the individual programmer,
- The new skills required to use the development chain are limited and compensated by a proportionate productivity increase,
- The development chain eases both design space exploration and parallel software/hardware development. Design space exploration is an early stage of system development consisting of testing algorithms on several architectures and making appropriate choices compromising between hardware and software optimisations, based on evaluated performances.
- The exploitation of the automatic parallelism of the development chain produces a nearly optimal result. For example, despite the impressive capabilities of the Texas Instruments TMS320C64x+ compiler, compute-hungry functions are still optimized by writing intrinsics or assembly code. Embedded multi-core development chains will only be adopted when programmers are no longer able to cope with efficient hand-parallelization.

There will always be a tradeoff between system `performance` and `programmability`; between system genericity and optimality. To be used, a development chain must connect with legacy code as well as easing design process. These principles were considered during the development of PREESM. PREESM plug-in functionalities are numerous and combined in graphical `workflows` that adapt the process to designer goals. PREESM clearly separates algorithm and architecture models to enable design space exploration, introducing an additional input entity named `scenario` that ensures this separation. The deployment of an algorithm on an architecture is automatic, as is the static code generation and the quality of a deployment is illustrated in a graphical `"schedule quality assessment chart"`. An important feature of PREESM is that a programmer can debug code on a single-core and then deploy it automatically over several cores with an assured absence of deadlocks.

However, there is a limitation to the compile-time deployment technique. If an algorithm is highly variable during its execution, choosing its execution configuration at compile-time is likely to bring excessive suboptimality. For the highly variable parts of an algorithm, the equivalent of an OS scheduler for multi-core architectures is thus needed. The resulting `adaptive scheduler` must be of very low complexity, manage architecture heterogeneity and substantially improve the resulting system quality.

Throughout this document, the idea of rapid prototyping and executable code generation from dataflow models is applied to the LTE physical layer algorithms.

1.2 Objectives

The aim of this work is to present and discuss efficient solutions for LTE deployment over heterogeneous multi-core architectures. A `fully tested method for rapid prototyping and automatic code generation` was developed from dataflow graphs. During the development, it became clear that there was a need for a new input entity or scenario to the rapid prototyping process . The scenario breaks the "Y" shape of the previous rapid prototyping methods and totally separates algorithm from architecture.

Existing architecture models appeared to be unable to describe the target architectures, so a novel architecture model is presented, the `System-Level Architecture Model` or S-LAM. This architecture model is intended to simplify the high-level view of an architecture as well as to accelerate the deployment.

Mimicing the ability of the SystemC Transaction Level Modeling (TLM) to offer scalability in the precision of target architecture simulations, a `scalable scheduler` was created, enabling tradeoffs between scheduling time and precision. A developer needs to evaluate the quality of a generated schedule and, more precisely, needs to know if the schedule parallelism is limited by the algorithm, by the architecture or by none of them. For this purpose, a literature-based, graphical `schedule quality assessment chart` is presented.

During the deployment of LTE algorithms, it became clear that, for these algorithms, using execution latency as the minimized criterion for scheduling did not produce good load balancing over the cores for the architectures studied. A new `scheduling criterion embedding latency and load balancing` was developed. This criterion leads to very balanced loads and, in the majority of cases, to an equivalent latency than simply using the latency criterion.

Finally, a `study of the LTE physical layer` in terms of rapid prototyping and code generation is presented. Some algorithms are too variable for simple compile-time scheduling, so an `adaptive scheduler` with the capacity to schedule the most dynamic algorithms of LTE at run-time was developed.

1.3 Outline

The outline of this document is depicted in Fig. 1.2. It is organized around the rapid prototyping and code generation process. After an introduction in Chap. 1.1, Part I presents elements from the literature used in Part II to create a rapid prototyping

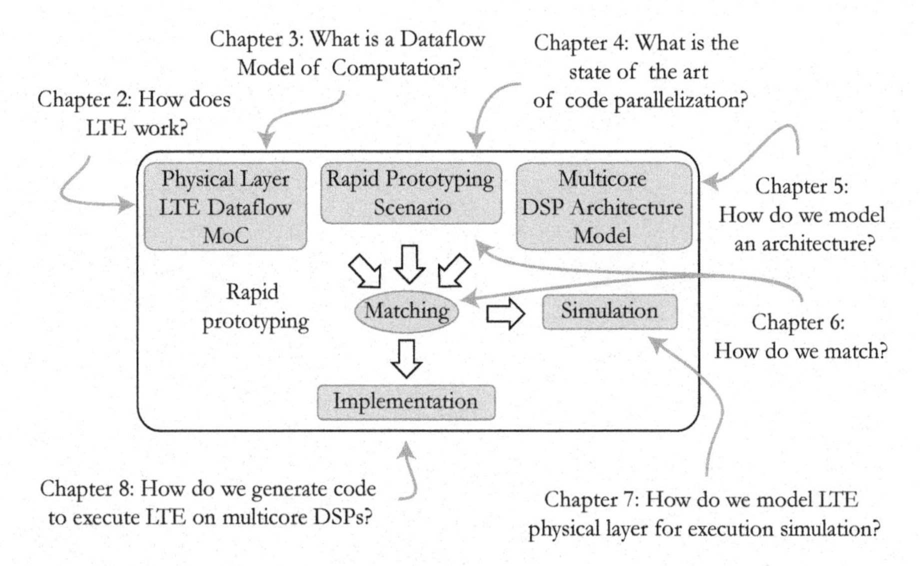

Fig. 1.2 Rapid prototyping and document outline

method which allows the study of LTE signal processing algorithms. In Chap. 2, the 3GPP LTE telecommunication standard is introduced. This chapter focuses on the signal processing features of LTE. In Chap. 3, the dataflow models of computation are explained; these are the models that are used to describe the algorithms in this study. Chap. 4 explains the existing techniques for system programming and rapid prototyping.

The S-LAM architecture model developed to feed the rapid prototyping method is presented in Chap. 5. In Chap. 6, a scheduler structure is detailed that separates the different problems of multi-core scheduling as well as some improvements to the state-of-the-art methods. The two last chapters are dedicated to the study of LTE and the application of all of the previously introduced techniques. Chapter 7 focuses on LTE rapid prototyping and simulation and Chap. 8 on the code generation. Chapter 8 is divided into two parts; the first dealing with static code generation, and the second with the heart of a multi-core operating system which enables dynamic code behavior: an adaptive scheduler. Chapter 9 concludes this study.

References

1. Ackerman WB (1982) Data flow languages computer. Data Flow Lang 15(2):15–25
2. Karam LJ, AlKamal I, Gatherer A, Frantz GA, Anderson DV, Evans BL (2009) Trends in multicore DSP platforms. IEEE Sig Process Mag 26(6):38–49

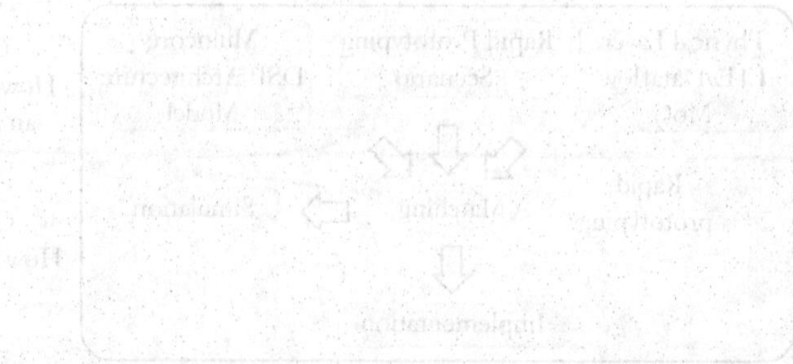

Chapter 2
3GPP Long Term Evolution

2.1 Introduction

2.1.1 Evolution and Environment of 3GPP Telecommunication Systems

Terrestrial mobile telecommunications started in the early 1980s using various analog systems developed in Japan and Europe. The Global System for Mobile communications (GSM) digital standard was subsequently developed by the European Telecommunications Standards Institute (ETSI) in the early 1990s. Available in 219 countries, GSM belongs to the second generation mobile phone system. It can provide an international mobility to its users by using inter-operator roaming. The success of GSM promoted the creation of the Third Generation Partnership Project (3GPP), a standard-developing organization dedicated to supporting GSM evolution and creating new telecommunication standards, in particular a Third Generation Telecommunication System (3G). The current members of 3GPP are ETSI (Europe), ATIS(USA), ARIB (Japan), TTC (Japan), CCSA (China) and TTA (Korea). In 2010, there are 1.3 million 2G and 3G base stations around the world [6] and the number of GSM users surpasses 3.5 billion [25].

The existence of multiple vendors and operators, the necessity interoperability when roaming and limited frequency resources justify the use of unified telecommunication standards such as GSM and 3G. Each decade, a new generation of standards multiplies the data rate available to its user by ten (Fig. 2.1). The driving force behind the creation of new standards is the radio spectrum which is an expensive resource shared by many interfering technologies. Spectrum use is coordinated by International Telecommunication Union, Radio Communication Sector (ITU-R), an international organization which defines technology families and assigns their spectral bands to frequencies that fit the International Mobile Telecommunications (IMT) requirements. 3G systems including LTE are referred to as ITU-R IMT-2000.

M. Pelcat et al., *Physical Layer Multi-Core Prototyping*,
Lecture Notes in Electrical Engineering 171, DOI: 10.1007/978-1-4471-4210-2_2,

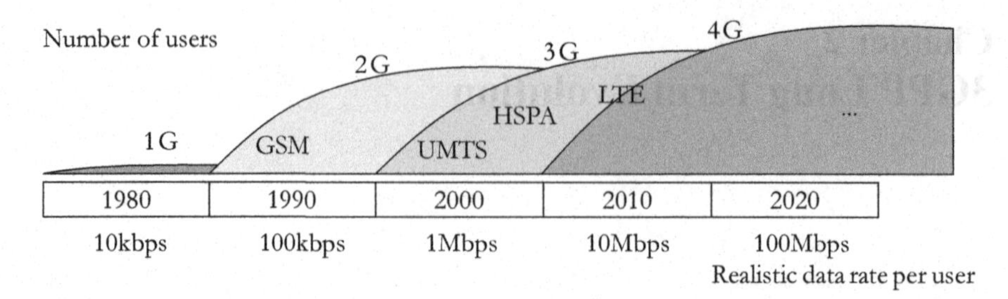

Fig. 2.1 3GPP standard generations

Radio access networks must constantly improve to accommodate the tremendous evolution of mobile electronic devices and internet services. Thus, 3GPP unceasingly updates its technologies and adds new standards. The goal of new standards is the improvement of key parameters, such as complexity, implementation cost and compatibility, with respect to earlier standards. Universal Mobile Telecommunications System (UMTS) is the first release of the 3G standard. Evolutions of UMTS such as High Speed Packet Access (HSPA), High Speed Packet Access Plus (HSPA+) or 3.5G have been released as standards due to providing increased data rates which enable new mobility internet services like television or high speed web browsing. The 3GPP Long Term Evolution (LTE) is the 3GPP standard released subsequent to HSPA+. It is designed to support the forecasted ten-fold growth of traffic per mobile between 2008 and 2015 [25] and the new dominance of internet data over voice in mobile systems. The LTE standardization process started in 2004 and a new enhancement of LTE named LTE-Advanced is currently being standardized.

2.1.2 Terminology and Requirements of LTE

A LTE terrestrial base station computational center is known as an evolved NodeB or `eNodeB`, where a NodeB is the name of a UMTS base station. An eNodeB can handle the communication of a few base stations, with each base station covering a geographic zone called a cell. A cell is usually three-sectored with three antennas (or antenna sets) each covering 120 (Fig. 2.2). The user mobile terminals (commonly mobile phones) are called User Equipment (`UE`). At any given time, a UE is located in one or more overlapping cells and communicates with a preferred cell; the one with the best air transmission properties. LTE is a duplex system, as communication flows in both directions between UEs and eNodeBs. The radio link between the eNodeB and the UE is called the downlink and the opposite link between UE and its eNodeB is called uplink. These links are asymmetric in data rates because most internet services necessitate a higher data rate for the downlink than for the uplink. Fortunately, it is easier to generate a higher data rate signal in an eNodeB powered by mains than in UE powered by batteries.

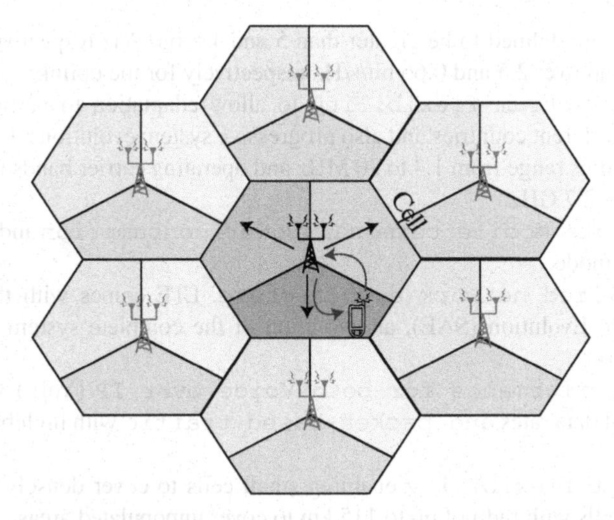

Fig. 2.2 A three-sectored cell

In GSM, UMTS and its evolutions, two different technologies are used for voice and data. Voice uses a circuit-switched technology, i.e. a resource is reserved for an active user throughout the entire communication, while data is packet-switched, i.e. data is encapsulated in packets allocated independently. Contrary to these predecessors, LTE is a totally packet-switched network using Internet Protocol (IP) and has no special physical features for voice communication. LTE is required to coexist with existing systems such as UMTS or HSPA in numerous frequency configurations and must be implemented without perturbing the existing networks.

LTE Radio Access Network advantages compared with previous standards (GSM, UMTS, HSPA...) are [30]:

- `Improved data rates`. Downlink peak rate are over 100 Mbit/s assuming 2 UE receive antennas and uplink peak rate over 50Mbit/s. Raw data rates are determined by $Bandwidth * Spectral\ Efficiency$ where the bandwidth (in Hz) is limited by the expensive frequency resource and ITU-R regulation and the spectral efficiency (in bit/s/Hz) is limited by emission power and channel capacity (Sect. 2.3.1). Within this raw data rate, a certain amount is used for control, and so is hidden from the user. In addition to peak data rates, LTE is designed to ensure a high system-level performance, delivering high data rates in real situations with average or poor radio conditions.
- A `reduced` data transmission `latency`. The two-way delay is under 10 ms.
- A `seamless mobility` with handover latency below 100 ms; handover is the transition when a given UE leaves one LTE cell to enter another one. 100 ms has been shown to be the maximal acceptable round trip delay for voice telephony of acceptable quality [30].
- `Reduced cost per bit`. This reduction occurs due to an improved spectral efficiency; spectrum is an expensive resource. Peak and average spectral

efficiencies are defined to be greater than 5 and 1.6 bit/s/Hz respectively for the downlink and over 2.5 and 0.66 bit/s/Hz respectively for the uplink.

- A `high spectrum flexibility` to allow adaptation to particular constraints of different countries and also progressive system evolutions. LTE operating bandwidths range from 1.4 to 20 MHz and operating carrier bands range from 698 MHz to 2.7 GHz.
- A `tolerable mobile terminal power consumption` and a very low power idle mode.
- A `simplified network architecture`. LTE comes with the System Architecture Evolution (SAE), an evolution of the complete system, including core network.
- A `good performance for both Voice over IP` (VoIP) with small but constant data rates and `packet-based traffic` with high but variable data rates.
- A `spatial flexibility` enabling small cells to cover densely populated areas and cells with radii of up to 115 km to cover unpopulated areas.
- The support of `high velocity UEs` with good performance up to 120 km/h and connectivity up to 350 km/h.
- The management of up to `200 active-state users per cell` of 5 MHz or less and 400 per cell of 10 MHz or more.

Depending on the type of UE (laptop, phone...), a tradeoff is found between data rate and UE memory and power consumption. LTE defines 5 UE categories supporting different LTE features and different data rates.

LTE also supports data broadcast (television for example) with a spectral efficiency over 1 bit/s/Hz. The broadcasted data cannot be handled like the user data because it is sent in real-time and must work in worst channel conditions without packet retransmission.

Both eNodeBs and UEs have emission power limitations in order to limit power consumption and protect public health. An outdoor eNodeB has a typical emission power of 40–46 dBm (10–40 W) depending on the configuration of the cell. An UE with power class 3 is limited to a peak transmission power of 23 dBm (200 mW). The standard allows for path-loss of roughly between 65 and 150 dB. This means that For 5 MHz bandwidth, a UE is able to receive data of power from -100 to -25 dBm (0.1 pW to 3.2 μW).

2.1.3 Scope and Organization of the LTE Study

The scope of this study is illustrated in Fig. 2.3. It concentrates on the Release 9 LTE physical layer in the eNodeB, i.e. the signal processing part of the LTE standard. 3GPP finalized the LTE Release 9 in December 2009. The physical layer (Open Systems Interconnection (OSI) layer 1) uplink and downlink baseband processing must share the eNodeB digital signal processing resources. The downlink baseband process is itself divided into channel coding that prepares the bit stream for transmission and

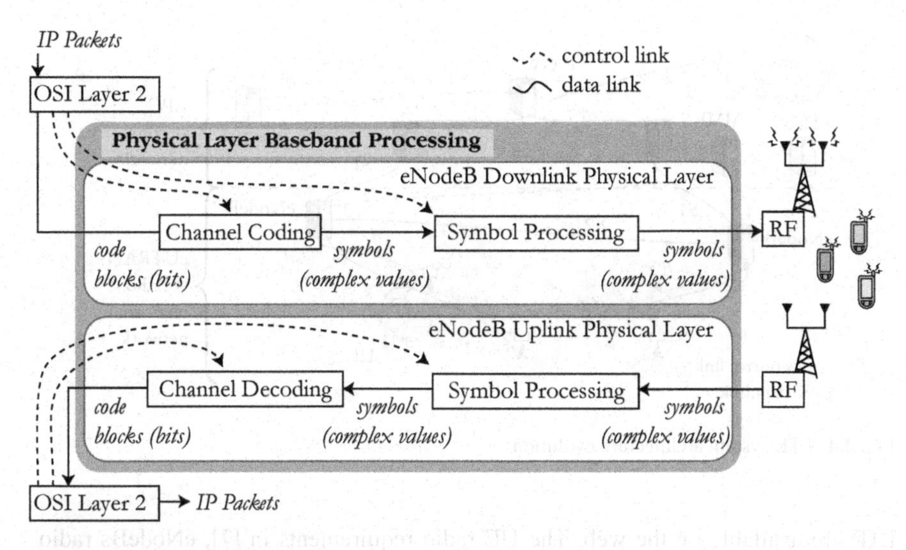

Fig. 2.3 Scope of the LTE study

symbol processing that adapts the signal to the transmission technology. The uplink baseband process performs the corresponding decoding. To explain the interaction with the physical layer, a short description of LTE network and higher layers will be given (in Sect. 2.2). The OSI layer 2 controls the physical layer parameters.

The goal of this study is to address the most computationally demanding use cases of LTE. Consequently, there is a particular focus on the highest bandwidth of 20 MHz for both the downlink and the uplink. An eNodeB can have up to 4 transmit and 4 receive antenna ports while a UE has 1 transmit and up to 2 receive antenna ports. An understanding of the basic physical layer functions assembled and prototyped in the rapid prototyping section is important. For this end, this study considers only the baseband signal processing of the physical layer. For transmission, this means a sequence of complex values $z(t) = x(t) + jy(t)$ used to modulate a carrier in phase and amplitude are generated from binary data and for each antenna port. A single antenna port carries a single complex value $s(t)$ at a one instant in time and can be connected to several antennas.

$$s(t) = x(t)\cos(2\pi f t) + y(t)\sin(2\pi f t) \tag{2.1}$$

where f is the carrier frequency which ranges from 698 MHz to 2.7GHz. The receiver gets an impaired version of the transmitted signal. The baseband receiver acquires complex values after lowpass filtering and sampling and reconstructing the transmitted data.

An overview of LTE OSI layers 1 and 2 with further details on physical layer technologies and their environment is presented in the following sections. A complete description of LTE can be found in [20, 21] and [30]. Standard documents describing

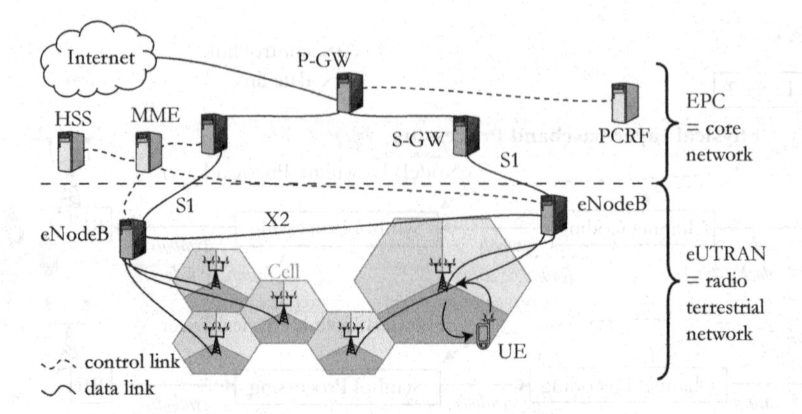

Fig. 2.4 LTE system architecture evolution

LTE are available on the web. The UE radio requirements in [7], eNodeBs radio requirements in [8], rules for uplink and downlink physical layer in [9] and channel coding in [10] with rules for defining the LTE physical layer.

2.2 From IP Packets to Air Transmission

2.2.1 Network Architecture

LTE introduces a new network architecture named System Architecture Evolution (SAE) and is displayed in Fig. 2.4 where control nodes are grayed compared with data nodes. SAE is divided into two parts:

- The Evolved Universal Terrestrial Radio Access Network (E-UTRAN) manages the radio resources and ensures the security of the transmitted data. It is composed entirely of eNodeBs. One eNodeB can manage several cells. Multiple eNodeBs are connected by cabled links called X2 allowing handover management between two close LTE cells. For the case where a handover occurs between two eNodeBs not connected by a X2 link, the procedure uses S1 links and is more complex.
- The Evolved Packet Core (EPC) also known as core network, enables packet communication with internet. The Serving Gateways (S-GW) and Packet Data Network Gateways (P-GW) ensure data transfers and Quality of Service (QoS) to the mobile UE. The Mobility Management Entities (MME) are scarce in the network. They handle the signaling between UE and EPC, including paging information, UE identity and location, communication security, load balancing. The radio-specific control information is called Access Stratum (AS). The radio-independent link between core network and UE is called Non-Access Stratum (NAS). MMEs delegate the verification of UE identities and operator subscriptions to Home Subscriber Servers (HSS). Policy Control and charging Rules

Function (PCRF) servers check that the QoS delivered to a UE is compatible with its subscription profile. For example, it can request limitations of the UE data rates because of specific subscription options.

The details of eNodeBs and their protocol stack are now described.

2.2.2 LTE Radio Link Protocol Layers

The information sent over a LTE radio link is divided in two categories: the `user-plane` which provides data and control information irrespective of LTE technology and the `control-plane` which gives control and signaling information for the LTE radio link. The protocol layers of LTE are displayed in Fig. 2.5 differ between user plane and control plane but the low layers are common to both planes. Figure 2.5 associates a unique OSI Reference Model number to each layer. layers 1 and 2 have identical functions in control-plane and user-plane even if parameters differ (for instance, the modulation constellation). Layers 1 and 2 are subdivided in:

- `PDCP layer` [14] or layer 2 Packet Data Convergence Protocol is responsible for data ciphering and IP header compression to reduce the IP header overhead. The service provided by PDCP to transfer IP packets is called a radio bearer. A radio bearer is defined as an IP stream corresponding to one service for one UE.
- `RLC layer` [13] or layer 2 Radio Link Control performs the data concatenation and then generates the segmentation of packets from IP-Packets of random sizes which comprise a Transport Block (TB) of size adapted to the radio transfer. The RLC layer also ensures ordered delivery of IP-Packets; Transport Block order can be modified by the radio link. Finally, the RLC layer handles a retransmission scheme of lost data through a first level of Automatic Repeat reQuests (ARQ). RLC manipulates logical channels that provide transfer abstraction services to the upper layer radio bearers. A radio bearer has a priority number and can have Guaranteed Bit Rate (GBR).
- `MAC layer` [12] or layer 2 Medium Access Control commands a low level retransmission scheme of lost data named Hybrid Automatic Repeat reQuest (HARQ). The MAC layer also multiplexes the RLC logical channels into HARQ protected transport channels for transmission to lower layers. Finally, the MAC layer contains the scheduler (Sect. 2.2.4), which is the primary decision maker for both downlink and uplink radio parameters.
- `Physical layer` [9] or layer 1 comprises all the radio technology required to transmit bits over the LTE radio link. This layer creates physical channels to carry information between eNodeBs and UEs and maps the MAC transport channels to these physical channels. The following sections focus on the physical layer with no distinction drawn between user and control planes.

`Layer 3` differs in control and user planes. Its Control plane handles all information specific to the radio technology, with the MME making the upper layer decisions. The User plane carries IP data from system end to system end (i.e. from UE to

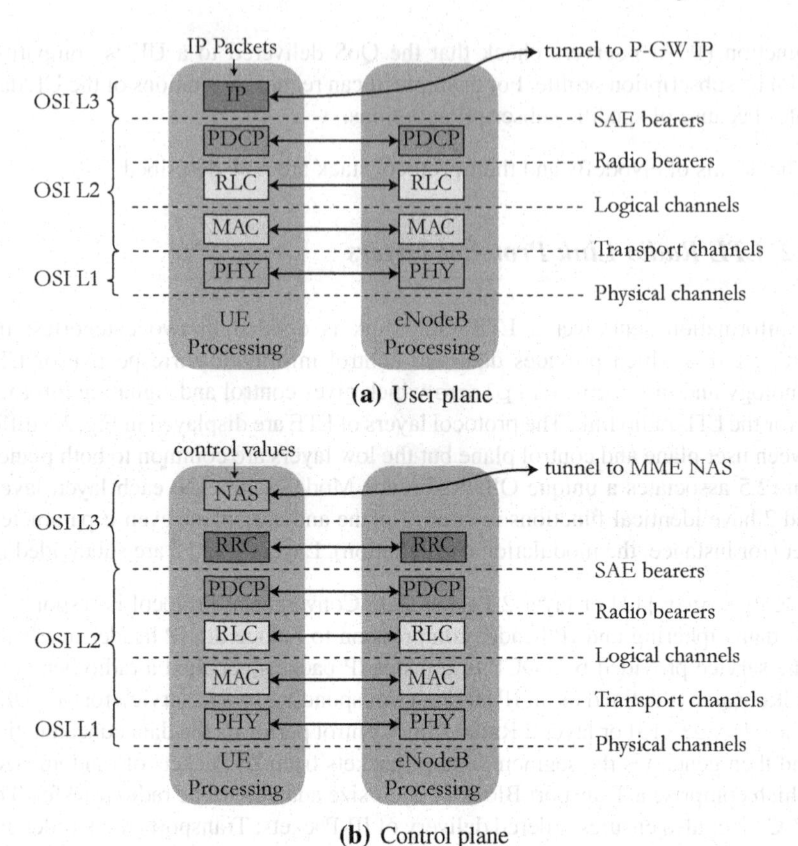

Fig. 2.5 Protocol layers of LTE radio link

P-GW). No further detail will be given on LTE non-physical layers. More information can be found in [20], p. 300 and [30], pp. 51 and 79.

Using both HARQ, employed for frequent and localized transmission errors, and ARQ, which is used for rare but lengthy transmission errors, results in high system reliability while limiting the error correction overhead. The retransmission in LTE is determined by the target service: LTE ensures different Qualities of Service (QoS) depending on the target service. For instance, the maximal LTE-allowed packet error loss rate is 10^{-2} for conversational voice and 10^{-6} for transfers based on Transmission Control Protocol (TCP) OSI layer 4. The various QoS imply different service priorities. For the example of a TCP/IP data transfer, the TCP packet retransmission system adds a third error correction system to the two LTE ARQs.

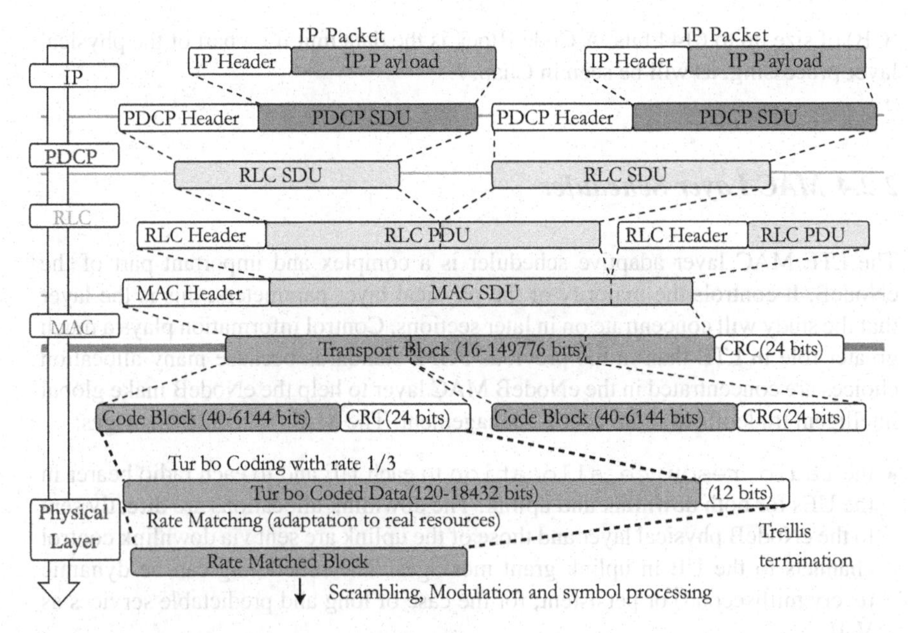

Fig. 2.6 Data blocks segmentation and concatenation

2.2.3 Data Blocks Segmentation and Concatenation

The physical layer manipulates bit sequences called Transport Blocks. In the user plane, many block segmentations and concatenations are processed layer after layer between the original data in IP packets and the data sent over air transmission. Figure 2.6 summarizes these block operations. Evidently, these operations do not reflect the entire bit transformation process including ciphering, retransmitting, ordering, and so on.

In the PDCP layer, the IP header is compressed and a new PDCP header is added to the ciphered Protocol Data Unit (PDU). In the RLC layer, RLC Service Data Units (SDU) are concatenated or segmented into RLC PDUs and a RLC header is added. The MAC layer concatenates RLC PDUs into MAC SDUs and adds a MAC header, forming a Transport Block, the data entity sent by the physical layer. For more details on layer 2 concatenation and segmentation, see [30], p. 79. The physical layer can carry downlink a Transport Blocks of size up to 149776 bits in 1 ms. This corresponds to a data rate of 149.776 Mbit/s. The overhead required by layer 2 and upper layers reduces this data rate. Moreover, such a Transport Block is only possible in very favorable transmission conditions with a UE capable of supporting the data rate. Transport Block sizes are determined from radio link adaptation parameters shown in the tables of [11], p. 26. An example of link capacity computing is given in Sect. 7.2.2. In the physical layer, Transport Blocks are segmented into Code Blocks

(CB) of size up to 6144 bits. A Code Block is the data unit for a part of the physical layer processing, as will be seen in Chap. 7.

2.2.4 MAC Layer Scheduler

The LTE MAC layer adaptive scheduler is a complex and important part of the eNodeB. It controls the majority of the physical layer parameters; this is the layer that the study will concentrate on in later sections. Control information plays a much greater role in LTE than in the previous 3GPP standards because many allocation choices are concentrated in the eNodeB MAC layer to help the eNodeB make global intelligent tradeoffs in radio access management. The MAC scheduler manages:

- the `radio resource allocation` to each UE and to each radio bearer in the UEs for both downlink and uplink. The downlink allocations are directly sent to the eNodeB physical layer and those of the uplink are sent via downlink control channels to the UE in uplink grant messages. The scheduling can be dynamic (every millisecond) or persistent, for the case of long and predictable services as VoIP.
- the `link adaptation` parameters (Sect. 2.3.5) for both downlink and uplink.
- the `HARQ` (Sect. 2.3.5.1) commands where lost Transport Blocks are retransmitted with new link adaptation parameters.
- the `Random Access Procedure` (Sect. 2.4.5) to connect UEs to a eNodeB.
- the `uplink timing alignment` (Sect. 2.3.4) to ensure UE messages do not overlap.

The MAC scheduler must take data priorities and properties into account before allocating resources. Scheduling also depends on the data buffering at both eNodeB and UE and on the transmission conditions for the given UE. The scheduling optimizes link performance depending on several metrics, including throughput, delay, spectral efficiency, and fairness between UEs.

2.3 Overview of LTE Physical Layer Technologies

2.3.1 Signal Air transmission and LTE

In [31], C. E. Shannon defines the capacity C of a communication channel impaired by an Additive White Gaussian Noise (AWGN) of power N as:

$$C = B \cdot \log_2(1 + \frac{S}{N}) \tag{2.2}$$

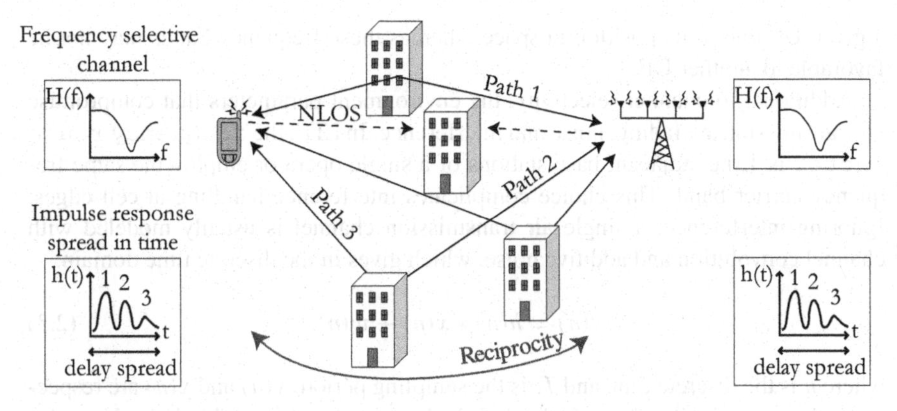

Fig. 2.7 Radio propagation, channel response and reciprocity property

where C is in bit/s and S is the signal received power. The `channel capacity` is thus linearly dependent on bandwidth. For the largest possible LTE bandwidth, 20 MHz, this corresponds to 133 Mbit/s or 6.65 bit/s/Hz for a $S/N = 20$ dB Signal-to-Noise Ratio or SNR (100 times more signal power than noise) and 8 Mbit/s or 0.4 bit/s/Hz for a -5 dB SNR (3 times more noise than signal). Augmenting the transmission power will result in an increased capacity, but this parameter is limited for security and energy consumption reasons. In LTE, the capacity can be doubled by creating two different channels via several antennas at transmitter and receiver sides. This technique is commonly called Multiple Input Multiple Output (`MIMO`) or spatial multiplexing and is limited by control signaling cost and non-null correlation between channels. It may be noted that the LTE target peak rate of 100 Mbit/s or 5 bit/s/Hz is close to the capacity of a single channel. Moreover, the real air transmission channel is far more complex than its AWGN model. Managing this complexity while maintaining data rates close to the channel capacity is one of the great challenges of LTE deployment.

LTE signals are transmitted from terrestrial base stations using electromagnetic waves propagating at light speed. LTE cells can have a radii of up to 115 km, leading to a transmission latency of about 380 μs in both downlink and uplink directions. The actual value of this latency depends on the cell radius and environment. Compensation of this propagation time is performed by UEs and called `timing advance` ([30], p. 459).

Moreover, the signal can undergo several reflections before reaching its target. This effect is known as `multiple path propagation` and is displayed in Fig. 2.7. In the time domain, multiple reflections create a Channel Impulse Response (CIR) $h(t)$ with several peaks, each corresponding to a reflection. This effect elongates a received symbol in time and can cause Inter Symbol Interference (ISI) between two successive symbols. The ISI introduces a variable attenuation over the frequency band generating a frequency selective channel. For a given time and a given cell, there are frequency bands highly favorable to data transmission between the eNodeB and a given UE due to its position in space whereas these frequency bands may not be favorable to another UE.

Additional to channel selectivity, the environment parameters that compromise air transmission are fading, noise and interference. In LTE, the `frequency reuse factor` is 1, i.e. adjacent base stations of a single operator employ the same frequency carrier band. This choice complicates interference handling at cell edges. Ignoring interference, a single air transmission channel is usually modeled with channel convolution and additive noise, which gives in the discrete time domain:

$$y(n) = h(n) * x(n) + w(n), \tag{2.3}$$

where n is the discrete time and T_s is the sampling period, $x(n)$ and $y(n)$ are respectively the transmitted and received signals in discrete time, $h(n)$ is the channel impulse response (Fig. 2.7) and $w(n)$ is the noise. The equivalent in Fourier discrete domain gives:

$$Y(k) = H(k)X(k) + W(k), \tag{2.4}$$

where k is the discrete frequency. In order to estimate $H(k)$, known reference signals (also called pilot signals) are transmitted. A reference signal cannot be received at the same time as the data it is aiding. Certain channel assumptions must be made, including slow modification over time. The time over which a Channel Impulse Response $h(t)$ remains almost constant is called `channel coherence time`. For a flat Rayleigh fading channel model at 2 GHz, modeling coherence time is about 5 ms for a UE speed of 20 km/h ([30], p. 576). The faster the UE moves, the faster the channel changes and the smaller the coherence time becomes.

The UE velocity also has an effect on radio propagation, due to the `Doppler effect`. For a carrier frequency of 2.5 GHz and a UE velocity of 130 km/h, the Doppler effect frequency shifts the signal up to 300 Hz ([30], p. 478). This frequency shift must be evaluated and compensated for each UE. Moreover, guard frequency bands between UEs are necessary to avoid frequency overlapping and Inter Carrier Interference (ICI).

Figure 2.7 shows a Non-line-of-sight (NLOS) channel, which occurs when the direct path is shadowed. Figure 2.7 also displays the property of `channel reciprocity`; the channels in downlink and in uplink can be considered to be equal in terms of frequency selectivity within the same frequency band. When downlink and uplink share the same band, channel reciprocity occurs, and so the uplink channel quality can be evaluated from downlink reception study and vice-versa. LTE technologies use channel property estimations $H(k)$ for two purposes:

- `Channel estimation` is used to reconstruct the transmitted signal from the received signal.
- `Channel sounding` is used by the eNodeBs to decide which resource to allocate to each UE. Special resources must be assigned to uplink channel soundings because a large frequency band exceeding UE resources must be sounded initially

by each UE to make efficient allocation decisions. The downlink channel sounding is quite straightforward, as the eNodeB sends reference signals over the entire downlink bandwidth.

Radio models describing several possible LTE environments have been developed by 3GPP (SCM and SCME models), ITU-R (IMT models) and a project named IST-WINNER. They offer tradeoffs between complexity and accuracy. Their targeted usage is hardware conformance tests. The models are of two kinds: matrix-based models simulate the propagation channel as a linear correlation (Eq. 2.3) while geometry-based models simulate the addition of several propagation paths (Fig. 2.7) and interferences between users and cells.

LTE is designed to address a variety of environments from mountainous to flat, including both rural and urban with Macro/Micro and Pico cells. On the other hand, Femtocells with very small radii are planned for deployment in indoor environments such as homes and small businesses. They are linked to the network via a Digital Subscriber Line (DSL) or cable.

2.3.2 Selective Channel Equalization

The air transmission channel attenuates each frequency differently, as seen in Fig. 2.7. Equalization at the decoder site consists of compensating for this effect and reconstructing the original signal as much as possible. For this purpose, the decoder must precisely evaluate the channel impulse response. The resulting coherent detection consists of 4 steps:

1. Each transmitting antenna sends a known `Reference Signal` (RS) using predefined time/frequency/space resources. Additional to their use for channel estimation, RS carry some control signal information. Reference signals are sometimes called pilot signals.
2. The RS is decoded and the $H(f)$ (Eq. 2.4) is computed for the RS time/frequency/space resources.
3. $H(f)$ is interpolated over time and frequency on the entire useful bandwidth.
4. Data is decoded exploiting $H(f)$.

The LTE uplink and downlink both exploit coherent detection but employ different reference signals. These signals are selected for their capacity to be orthogonal with each other and to be detectable when impaired by Doppler or multipath effect. Orthogonality implies that several different reference signals can be sent by the same resource and still be detectable. This effect is called Code Division Multiplexing (CDM). Reference signals are chosen to have constant amplitude, reducing the transmitted Peak to Average Power Ratio (PAPR, [28]) and augmenting the transmission power efficiency. Uplink reference signals will be explained in 2.4.3 and downlink reference signals in 2.5.3.

As the transmitted reference signal $X_p(k)$ is known at transmitter and receiver, it can be localized and detected. The simplest least square estimation defines:

$$H(k) = (Y(k) - W(k))/X_p(k) \approx Y(k)/X_p(k). \quad (2.5)$$

$H(k)$ can be interpolated for non-RS resources, by considering that channel coherence is high between RS locations. The transmitted data is then reconstructed in the Fourier domain with $X(k) = Y(k)/H(k)$.

2.3.3 eNodeB Physical Layer Data Processing

Figure 2.8 provides more details of the eNodeB physical layer that was roughly described in Fig. 2.3. It is still a simplified view of the physical layer that will be explained in the next sections and modeled in Chap. 7.

In the `downlink data encoding`, `channel coding` (also named link adaptation) prepares the binary information for transmission. It consists in a Cyclic Redundancy Check (CRC) /turbo coding phase that processes Forward Error Correction (FEC), a rate matching phase to introduce the necessary amount of redundancy, a scrambling phase to increase the signal robustness, and a modulation phase that transforms the bits into symbols. The parameters of channel coding are named Modulation and Coding Scheme (MCS). They are detailed in Sect. 2.3.5. After channel coding, `symbol processing` prepares the data for transmission over several antennas and subcarriers. The downlink transmission schemes with multiple antennas are explained in Sects. 2.3.6 and 2.5.4 and the Orthogonal Frequency Division Multiplexing Access (OFDMA), that allocates data to subcarriers, in Sect. 2.3.4.

In the `uplink data decoding`, the `symbol processing` consists in decoding Single Carrier-Frequency Division Multiplexing Access (SC-FDMA) and equalizing signals from the different antennas using channel estimates. SC-FDMA is the uplink broadband transmission technology and is presented in Sect. 2.3.4. Uplink multiple antenna transmission schemes are explained in Sect. 2.4.4. After symbol processing, uplink `channel decoding` consists of the inverse phases of downlink channel coding because the chosen techniques are equivalent to the ones of downlink. HARQ combining associates the repeated receptions of a single block to increase robustness in case of transmission errors.

Next sections explain in details these features of the eNodeB physical layer, starting with the broadband technologies.

2.3.4 Multicarrier Broadband Technologies and Resources

LTE uplink and downlink data streams are illustrated in Fig. 2.9. The LTE uplink and downlink both employ technologies that enable a two-dimension allocation of resources to UEs in time and frequency. A third dimension in space is added by Multiple Input Multiple Output (MIMO) spatial multiplexing (Sect. 2.3.6). The eNodeB decides the allocation for both downlink and uplink. The uplink allocation decisions

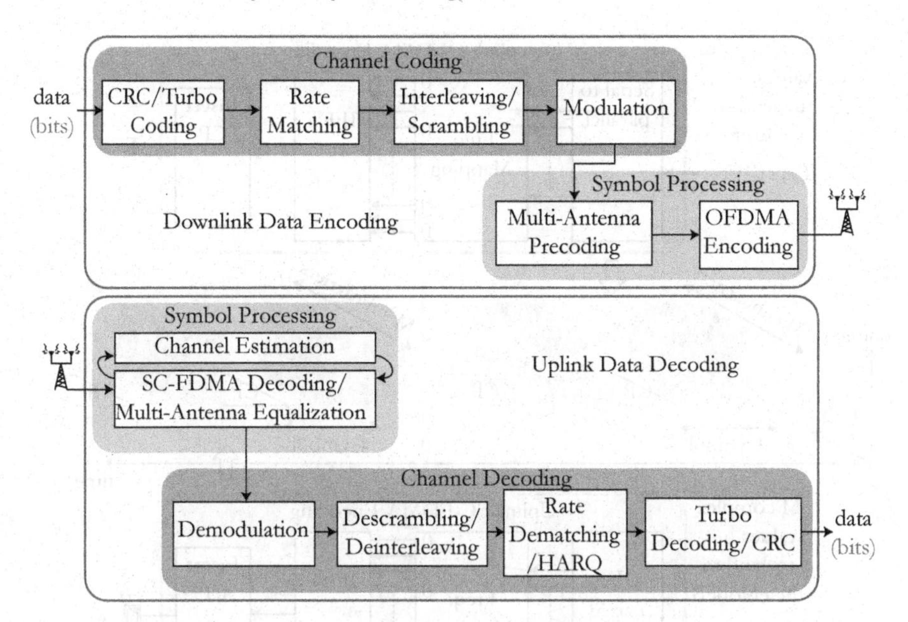

Fig. 2.8 Uplink and downlink data processing in the LTE eNodeB

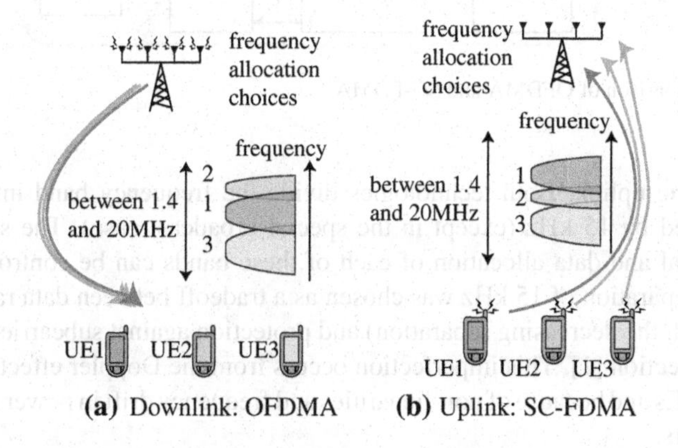

Fig. 2.9 LTE downlink and uplink multiplexing technologies

must be sent via the downlink control channels. Both downlink and uplink bands have six possible bandwidths: 1.4, 3, 5, 10, 15, or 20 MHz.

2.3.4.1 Broadband Technologies

The multiple subcarrier broadband technologies used in LTE are illustrated in Fig. 2.10. Orthogonal Frequency Division Multiplexing Access (OFDMA) employed for the downlink and Single Carrier-Frequency Division Multiplexing (SC-FDMA)

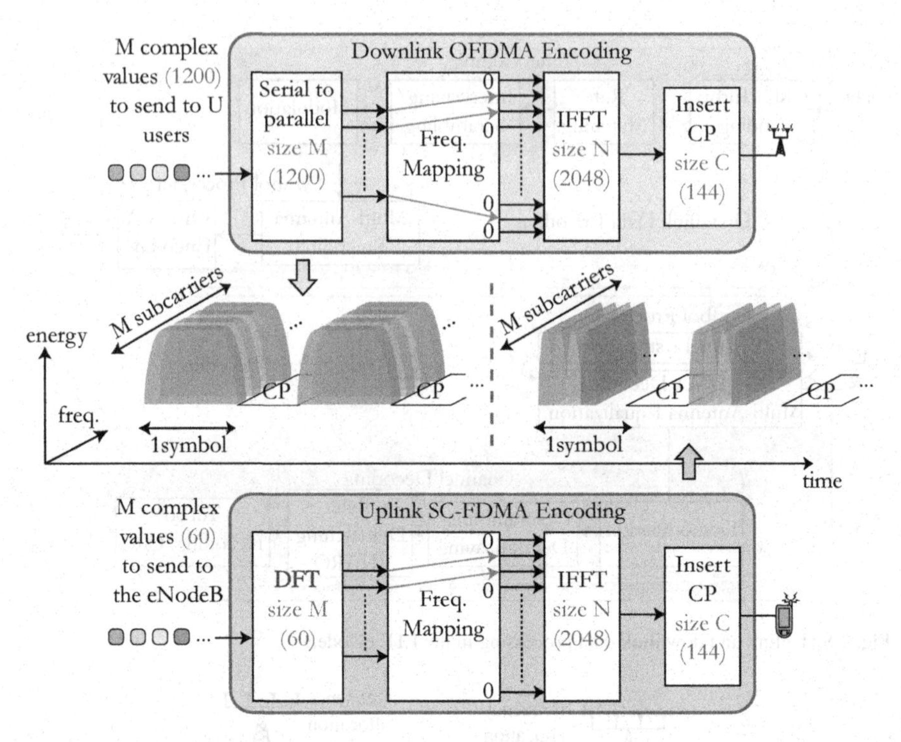

Fig. 2.10 Comparison of OFDMA and SC-FDMA

is used for the uplink. Both technologies divide the frequency band into subcarriers separated by 15 kHz (except in the special broadcast case). The subcarriers are orthogonal and data allocation of each of these bands can be controlled separately. The separation of 15 kHz was chosen as a tradeoff between data rate (which increases with the decreasing separation) and protection against subcarrier orthogonality imperfection [1]. This imperfection occurs from the Doppler effect produced by moving UEs and because of non-linearities and frequency drift in power amplifiers and oscillators.

Both technologies are effective in limiting the impact of multi-path propagation on data rate. Moreover, the dividing the spectrum into subcarriers enables simultaneous access to UEs in different frequency bands. However, SC-FDMA is more efficient than OFDMA in terms of Peak to Average Power Ratio (PAPR, [28]). The lower PAPR lowers the cost of the UE RF transmitter but SC-FDMA cannot support data rates as high as OFDMA in frequency-selective environments.

Figure 2.10 shows typical transmitter implementations of OFDMA and SC-FDMA using Fourier transforms. SC-FDMA can be interpreted as a linearly precoded OFDMA scheme, in the sense that it has an additional DFT processing preceding the conventional OFDMA processing. The frequency mapping of Fig. 2.10 defines the subcarrier accessed by a given UE.

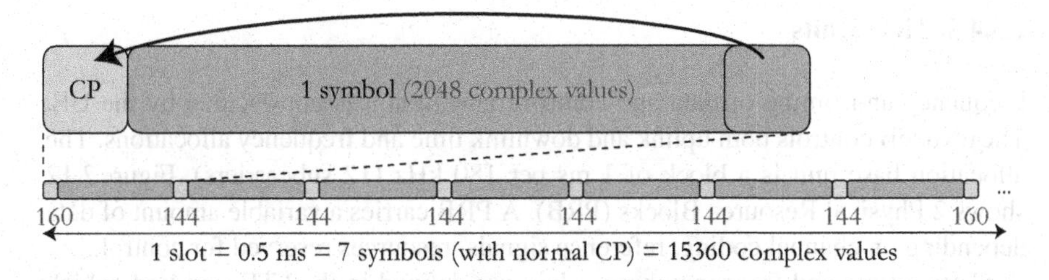

Fig. 2.11 Cyclic prefix insertion

Downlink symbol processing consists of mapping input values to subcarriers by performing an Inverse Fast Fourier Transform (IFFT). Each complex value is then transmitted on a single subcarrier but spread over an entire symbol in time. This transmission scheme protects the signal from Inter Symbol Interference (ISI) due to multipath transmission. It is important to note that without channel coding (i.e. data redundancy and data spreading over several subcarriers), the signal would be vulnerable to frequency selective channels and Inter Carrier Interference (ICI). The numbers in gray (in Fig. 2.10) reflect typical parameter values for a signal of bandwidth of 20 MHz. The OFDMA encoding is processed in the eNodeB and the 1200 input values of the case of 20MHz bandwidth carry the data of all the addressed UEs. SC-FDMA consists of a small size Discrete Fourier Transform (DFT) followed by OFDMA processing. The small size of the DFT is required as this processing is performed within a UE and only uses the data of this UE. For an example of 60 complex values the UE will use 60 subcarriers of the spectrum (subcarriers are shown later to be grouped by 12). As noted before, without channel coding, data would be prone to errors introduced by the wireless channel conditions, especially because of ISI in the SC-FDMA case.

2.3.4.2 Cyclic Prefix

The Cyclic Prefix (CP) displayed in Figs. 2.10 and 2.11 is used to separate two successive symbols and thus reduces ISI. The CP is copied from the end of the symbol data to an empty time slot reserved before the symbol and protects the received data from timing advance errors; the linear convolution of the data with the channel impulse response is converted into a circular convolution, making it equivalent to a Fourier domain multiplication that can be equalized after a channel estimation (Sect. 2.3.2). CP length in LTE is 144 $samples = 4.8\,\mu s$ (normal CP) or 512 $samples = 16.7\,\mu$s in large cells (extended CP). A longer CP can be used for broadcast when all eNodeBs transfer the same data on the same resources, so introducing a potentially rich multi-path channel. Generally, multipath propagation can be seen to induce channel impulse responses longer than CP. The CP length is a tradeoff between the CP overhead and sufficient ISI cancellation [1].

2.3.4.3 Time Units

Frequency and timing of data and control transmission is not decided by the UE. The eNodeB controls both uplink and downlink time and frequency allocations. The allocation base unit is a block of 1 ms per 180 kHz (12 subcarriers). Figure 2.12 shows 2 Physical Resource Blocks (PRB). A PRB carries a variable amount of data depending on channel coding, reference signals, resources reserved for control...

Certain time and frequency base values are defined in the LTE standard, which allows devices from different companies to interconnect flawlessly. The LTE time units are displayed in Fig. 2.12:

- A `basic time unit` lasts $T_s = 1/30720000\,\mathrm{s} \approx 33\,\mathrm{ns}$. This is the duration of 1 complex sample in the case of 20 MHz bandwidth. The sampling frequency is thus $30.72\,\mathrm{MHz} = 8 * 3.84\,\mathrm{MHz}$, eight times the sampling frequency of UMTS. The choice was made to simplify the RF chain used commonly for UMTS and LTE. Moreover, as classic OFDMA and SC-FDMA processing uses Fourier transforms, symbols of size power of two enable the use of FFTs and $30.72\,\mathrm{MHz} = 2048 * 15\,\mathrm{kHz} = 2^{11} * 15\,\mathrm{kHz}$, with 15 kHz the size of a subcarrier and 2048 a power of two. Time duration for all other time parameters in LTE is a multiple of T_s.
- A `slot` is of length $0.5\,\mathrm{ms} = 15360\,T_s$. This is also the time length of a PRB. A slot contains 7 symbols in normal cyclic prefix case and 6 symbols in extended CP case. A Resource Element (RE) is a little element of 1 subcarrier per one symbol.
- A `subframe` lasts $1\,\mathrm{ms} = 30720\,T_s = 2\,slots$. This is the minimum duration that can be allocated to a user in downlink or uplink. A subframe is also called Transmission Time Interval (TTI) as it is the minimum duration of an independently decodable transmission. A subframe contains 14 symbols with normal cyclic prefix that are indexed from 0 to 13 and are described in the following sections.
- A `frame` lasts $10\,\mathrm{ms} = 307200\,T_s$. This corresponds to the time required to repeat a resource allocation pattern separating uplink and downlink in time in case of Time Division Duplex (TDD) mode. TDD is defined below.

In the LTE standard, the subframe size of 1 ms was chosen as a tradeoff between a short subframe which introduces high control overhead and a long subframe which significantly increases the retransmission latency when packets are lost [3]. Depending on the assigned bandwidth, an LTE cell can have between 6 and 100 resource blocks per slot. In TDD, special subframes protect uplink and downlink signals from ISI by introducing a guard period ([9], p. 9).

2.3.4.4 Duplex Modes

Figure 2.13 shows the duplex modes available for LTE. Duplex modes define how the downlink and uplink bands are allocated respective to each other. In Frequency Division Duplex (`FDD`) mode, the uplink and downlink bands are disjoint. The connection is then full duplex and the UE needs to have two distinct Radio Frequency (RF) processing chains for transmission and reception. In Time Division

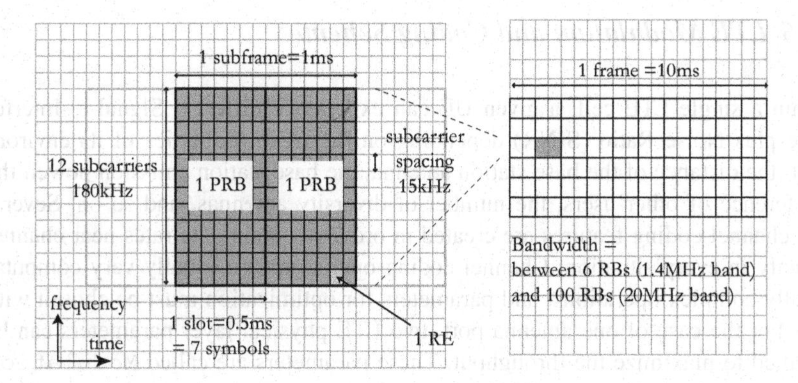

Fig. 2.12 LTE time units

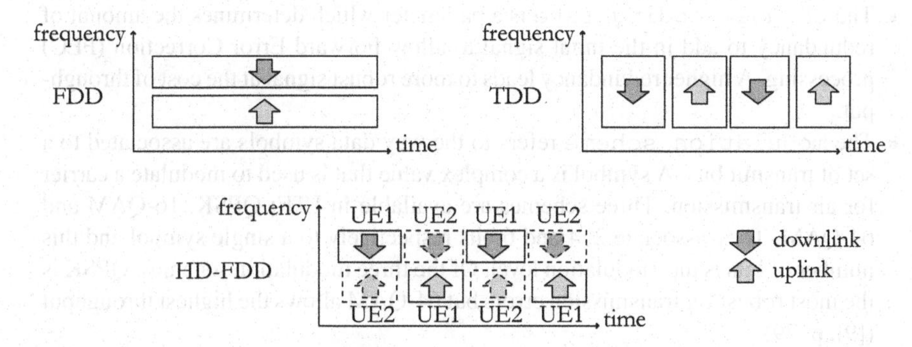

Fig. 2.13 Different types of LTE duplex modes

Duplex (TDD) mode, the downlink and the uplink alternatively occupy the same frequency band. The same RF chain can then be used for transmitting and receiving but available resources are halved. In Half-Duplex FDD (HD-FDD) mode, the eNodeB is full duplex but the UE is half-duplex (so can have a single RF chain). In this mode, separate bands are used for the uplink and the downlink but are never simultaneous for a given UE. HD-FDD is already present in GSM.

ITU-R defined 17 FDD and 8 TDD frequency bands shared by LTE and UMTS standards. These bands are located between 698 and 2690 MHz and lead to very different channel behavior depending on carrier frequency. These differences must be accounted for during the calibration of base stations and UEs. The following Sections will focus on FDD mode.

2.3.5 LTE Modulation and Coding Scheme

Within a single LTE cell, a given UE can experience different Signal-to-Interference plus Noise Ratio (SINR) depending on the radio properties of its environment: the distance of the base station antenna, the base station emission power, the interference of other users, the number of diversity antennas, and so on. Several LTE channel coding features are created in order to obtain data rates near channel capacity in every situation. Channel coding operations are usually very computationally complex operations, and parameters for optimization must be chosen with care. For the case of one antenna port, two LTE physical layer parameters can be modified to maximize the throughput. These parameters are called `Modulation and Coding Scheme` (MCS):

- The `channel coding rate` is a parameter which determines the amount of redundancy to add in the input signal to allow Forward Error Correction (FEC) processing. A higher redundancy leads to more robust signal at the cost of throughput.
- The `modulation scheme` refers to the way data symbols are associated to a set of transmit bits. A symbol is a complex value that is used to modulate a carrier for air transmission. Three schemes are available in LTE: QPSK, 16-QAM and 64-QAM. They associate 2, 4 and 6 bits respectively to a single symbol and this number of bits is the modulation level. Of the three modulation schemes, QPSK is the most robust for transmission errors but 64-QAM allows the highest throughput ([9], p. 79).

The downlink channel quality estimation required for downlink MCS scheme choice is more complex than for its uplink equivalent. However, the report of downlink channel quality is vital for the eNodeB when making downlink scheduling decisions. In FDD mode, no reciprocity of frequency selective fading between uplink and downlink channels can be used (Sect. 2.3.1). The UE measures downlink channel quality from downlink reference signals and then reports this information to its eNodeB. The UE report consists of a number between 0 and 15, generated by the UE, representing the channel capacity for a given bandwidth. This number is called CQI for `Channel Quality Indicator` and is sent to the eNodeB in the uplink control channel. The CQI influences the choice of resource allocation and MCS scheme. In Fig. 2.14a, the channel coding rate and modulation rate are plotted against CQI, and the global resulting coding efficiency. It may be seen that the coding efficiency can be gradually adapted from a transmission rate of 0.15 bits/resource element to 5.55 bits/resource element.

CQI reports are periodic, unless the eNodeB explicitly requests aperiodic reports. Periodic reports pass through the main uplink control channel known as the Physical Uplink Control CHannel (PUCCH, Sect. 2.4.2). When PUCCH resources are unavailable, the reports are multiplexed in the uplink data channel known as the Physical Uplink Shared CHannel (PUSCH, Sect. 2.4.2). Aperiodic reports are sent in the PUSCH when explicitly requested by the eNodeB. Periodic CQI reports have

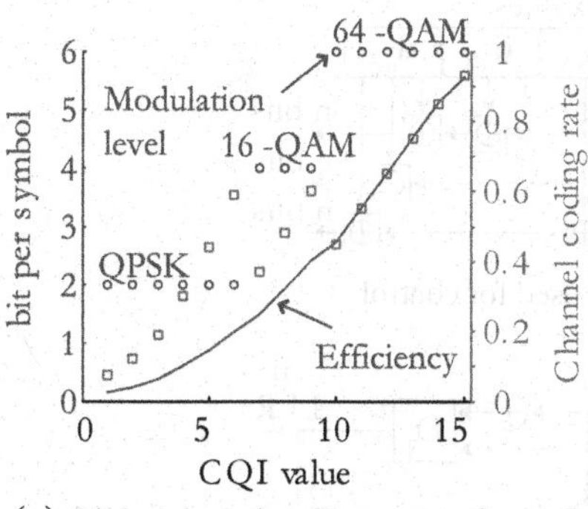

(a) Link Adaptation Parameters for each CQI

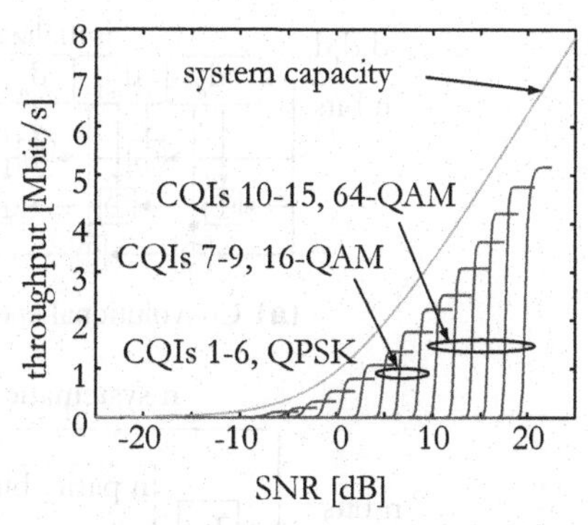

(b) Link Adaptation Effect on the Throughput of a 1.4 MHz Cell

Fig. 2.14 LTE link adaptation

a period between 2 and 160 ms. Periodic CQI reports contain one CQI if no spatial multiplexing is used or two CQI (one per rank) in the case of rank 2 downlink spatial multiplexing (Sect. 2.3.6). Aperiodic CQI modes exist including one CQI for the whole band and possibly an additional CQI for a set of preferred subbands ([11], p. 37). The choice of following the UE recommendation is given to the eNodeB. After receiving the UE report, the eNodeB sends data using the downlink data channel known as the Physical Downlink Shared Channel (PDSCH, Sect. 2.5.2). Control values are simultaneously transmitting in the main downlink control channel known as the Physical Downlink Control Channel (PDCCH, Sect. 2.5.2). These PDCCH control values carry Downlink Control Information (DCI) which include the chosen MCS scheme and HARQ parameters.

Figure 2.14b shows the effect of MCS on the throughput of a LTE transmission using a single antenna, a bandwidth of 1.4MHz, no HARQ and one UE. With only one UE in the cell, there can be no interference, so SINR is equal to SNR (Signal-to-Noise Ratio). The results were generated by the LTE simulator of Vienna University of Technology with a Additive White Gaussian Noise (AWGN) channel model and published in [24]. It may be seen that, for the studied channel model, a throughput close to the channel capacity can be achieved if the link adaptation is well chosen. It may also be noted that this choice is very sensitive to the SNR.

The techniques used to add redundancy and process Forward Error Correction (FEC) are different for control and data channels. The majority of control information in the PDCCH and the PUCCH use tail biting convolutional coding, sequence repetition and pruning while data channels (PDSCH, PUSCH) use turbo coding, sequence repetition and pruning ([29], p. 76).

The `convolutional code` used for control in LTE (Fig. 2.15a) has a 1/3 rate i.e. the code adds 2 redundant bits for each information bit. The encoder was chosen for its simplicity to allow to easier PDCCH decoding by UEs. Indeed, a UE needs to permanently decode many PDCCH PRBs, including many that are not intended for

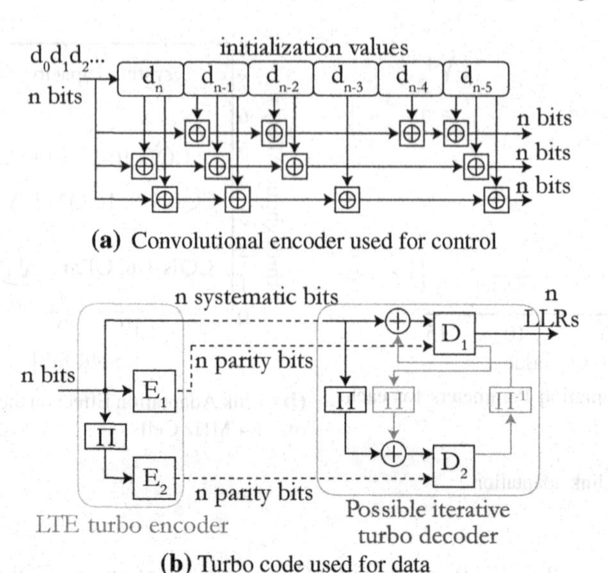

Fig. 2.15 LTE forward error correction methods

decoding and so this permanent computation must be limited. Convolutional coding is well suited for a FEC system with small blocks because it is not necessary to start in a predefined state. Moreover, convolutional codes can be efficiently decoded using a Viterbi decoder [19]. The absence of predefined state is important because a predefined starting state has a cost in the encoded stream. Instead, the 6 last bits of the encoded sequence serve as starting state of the encoder, transforming a linear convolution into a circular convolution, in the same way as the Cyclic Prefix with the channel impulse response. This technique is named tail biting.

A `turbo code` [16] in LTE (Fig. 2.15b) introduces redundancy in all LTE transmitted data. It has a rate of 1/3 and consists of an interleaver and two identical rate-1 convolutional encoders E_1 and E_2. One convolutional encoder (E_1) processes the input data while the other convolutional encoder (E_2) processes a pseudo-randomly interleaved version of the same data. The turbo coder outputs a concatenation of its unmodified input (systematic bits) and the two convoluted signals (parity bits). The resulting output facilitates FEC because there is little probability of having a low Hamming weight (number of 1s in the bit stream), thus improving detection. Each Code Block is independently turbo coded. A Code Block size ranges from 40 to 6144 bits (Sect. 2.2.3). Turbo coding is very efficient for long Code Blocks but it necessitates 12 tail bits containing no information that reduce its efficiency for small blocks.

Turbo coding and decoding are two of the most demanding functions of the LTE physical layer. Special coprocessors for turbo coding and decoding are included in multi-core architectures intended to handle eNodeB physical layer computation. Turbo iterative decoders ([27] and Fig. 2.15b) contain two convolution decoders that calculate the A Posteriori Probability (APP) of their input. At each iteration of the decoder, the detection belief of the decoded bit increases. APPs are stored in Log Likelihood Ratio (LLR) integer values where the sign indicates the detected binary value (0 or 1) and the amplitude indicates the reliability of the detection. For each bit, the turbo decoder outputs an LLR. Such a decoder, manipulating not only raw bits but also detection belief, is called soft decision decoder. For the possibility of HARQ retransmissions, LLR values of preceding receptions are stored to allow the combination of the preceding with the new reception. This combining operation can increase the signal to noise ratio in a code block and hence increase the odds for a successful decoding by the Turbo decoder.

After convolutional or turbo coding, the bits enter a circular Rate Matching (RM) process where they are interlaced, repeated and pruned to obtain the desired coding rate (Fig. 2.14a) with an optimal amount of spread redundancy.

2.3.5.1 Hybrid Automatic Repeat ReQuest

HARQ retransmission of lost blocks is a part of link adaptation. HARQ introduces redundancy in the signal to counteract the channel impairments. Moreover, HARQ is `hybrid` in the sense that each retransmission can be made more robust by a stronger modulation and by stronger channel coding.

An eNodeB has 3 ms after the end of a PUSCH subframe reception to process the frame, detect a possible reception error via Cyclic Redundancy Check (CRC) decoding and send back a NACK bit in the PDCCH. The uplink HARQ is synchronous in that a retransmission, if any, always appears 8 ms after the end of the first transmission. This fixed retransmission scheme reduces signaling in PUCCH. The repetitions can be adaptive (each one with a different MCS scheme), or not.

Downlink HARQ is consistently asynchronous and adaptive. There is at least 8 ms between two retransmissions of a Transport Block. It introduces more flexibility than the uplink scheme at the cost of signaling.

When a Transport Block is not correctly received, 8 different stop-and-wait processes are enabled, for both downlink and uplink receivers. This number of processes reduces the risk of the communication being blocked while waiting for HARQ acknowledgement, with the cost of memory to store the LLRs for each process (Sect. 2.3.5). Data from several repetitions are stored as LLRs and recombined, making HARQ hybrid.

2.3.6 Multiple Antennas

In LTE, eNodeBs and UEs can use several antennas to improve the quality of wireless links:

- eNodeB baseband processing can generate up to 4 separate downlink signals. Each signal is allocated to an antenna port. The eNodeB can also receive up to 4 different uplink signals. Two antennas sending the same signal are processed at the baseband processing as a single antenna port.
- A UE can have 2 receive antennas and receive 2 different signals simultaneously. It can have 2 transmit antennas but the UE will switch between these antennas as it has only one RF amplifier.

Multiple transmit antennas ($N_T > 1$), receive antennas ($N_R > 1$) or both can improve link quality and lead to higher data rates with higher spectral efficiency. For a precise introduction to MIMO channel capacity, see [22]. Different multiple antenna effects can be combined:

- `Spatial diversity` (Fig. 2.16a) consists of using different paths between antenna sets to compensate for the selective frequency fading of channels due to multipath transmission. There are two modes of spatial diversity: transmission or reception. Transmission spatial diversity necessitates several transmit antennas. Combining several transmit antennas consists in choosing the right way to distribute data over several antennas. A common technique is named Space-Time Block Coding (STBC) where successive data streams are multiplexed in the channels and made as orthogonal as possible to enhance the reception diversity. The most common of these codes is Alamouti; it offers optimal orthogonality for 2 transmit antennas [15]. No such optimal code exists for more than 2 antennas. Reception spatial diversity (Fig. 2.16a) uses several receive antennas exploiting the diverse channels, by combining their signal with Maximal-Ratio Combining (MRC).
- `Spatial multiplexing gain`, sometimes called MIMO effect is displayed in Fig. 2.16b for the case $N_T = N_R = 2$. It consists of creating several independent channels in space, exploiting a good knowledge of the channels and reconstructing the original data of each channel. MIMO can increase the throughput by a factor of $\min(N_T, N_R)$ but this factor is never obtained in real life because multiple paths are never totally uncorrelated and channel estimation requires additional reference signals. MIMO actually takes advantage of the multipath scattering that spatial diversity tries to counteract.
- `Beamforming`, also called array gain, consists of creating constructive and destructive interference of the wavefront to concentrate the signal in a given spatial direction. This is displayed in Fig. 2.16c. These interferences are generated by several transmitting antennas precoded by factors named beam patterns.

Without spatial multiplexing, every PRB corresponding to a single UE has the same MCS. In downlink spatial multiplexing mode, two separate MCS can be used for two different Transport Blocks sent simultaneously to a UE. A more complex scheme was shown to greatly increase the control with only slightly higher data rates [2].

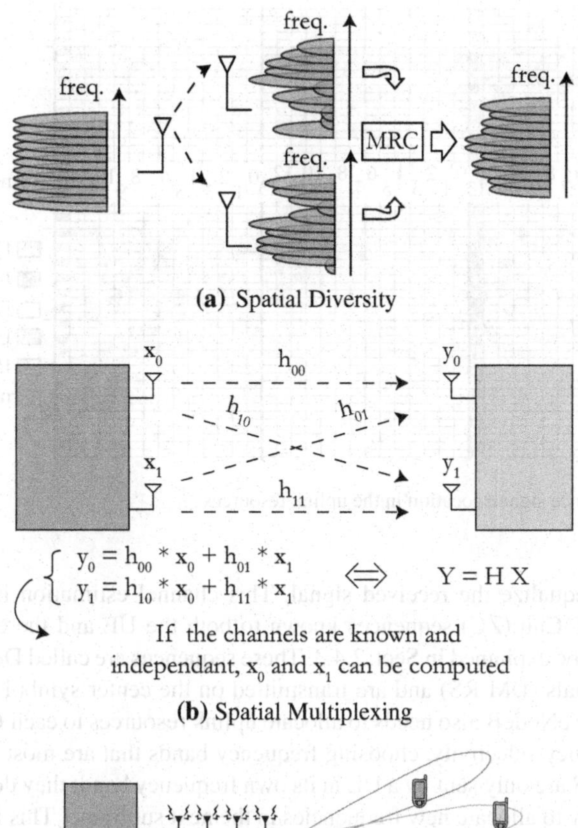

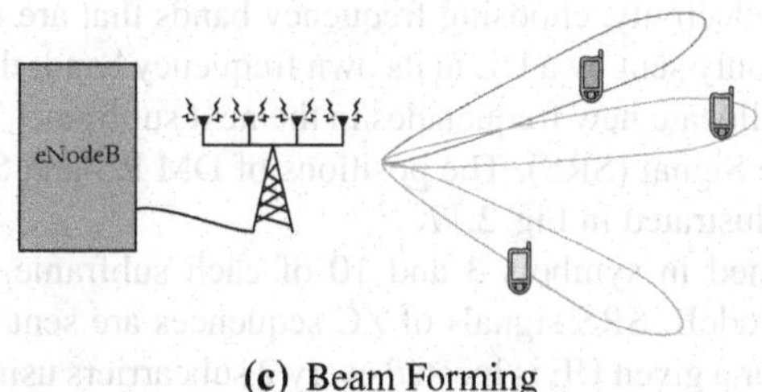

Fig. 2.16 LTE link adaptation

2.4 LTE Uplink Features

2.4.1 Single Carrier-Frequency Division Multiplexing

2.4.1.1 Uplink Pilot Signals

The SC-FDMA presented earlier is also known as DFT-Spread Orthogonal Frequency Division Multiplexing (DFTS-OFDM). A discussion on uplink technology choices can be found in [18]. A SC-FDMA decoder in the eNodeB necessitates a channel

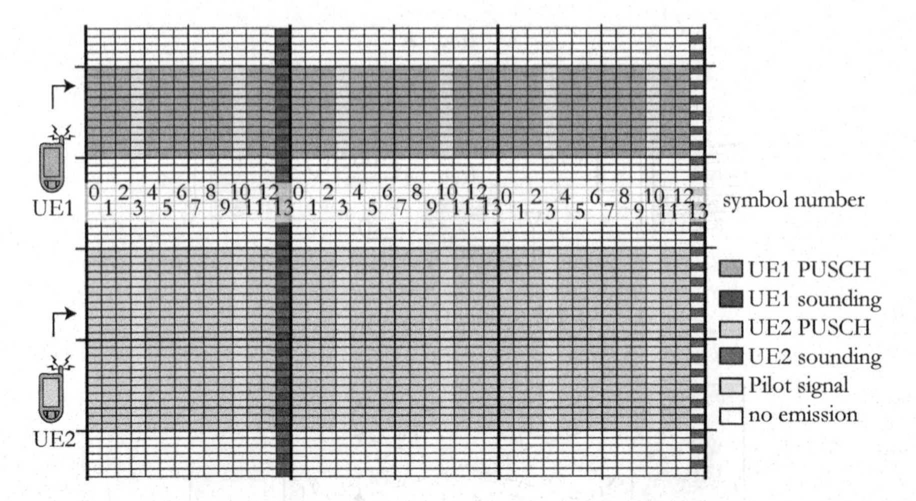

Fig. 2.17 Reference signals location in the uplink resources

estimation to equalize the received signal. This channel estimation is performed sending Zadoff–Chu (ZC) sequences known to both the UE and the eNodeB. ZC sequences will be explained in Sect. 2.4.3. These sequences are called Demodulation Reference Signals (DM RS) and are transmitted on the center symbol of each slot (Fig. 2.17). The eNodeB also needs to allocate uplink resources to each UE based on channel frequency selectivity, choosing frequency bands that are most favorable to the UE. DM RS are only sent by a UE in its own frequency band; they do not predict if it is necessary to allocate new frequencies in the next subframe. This is the role of Sounding Reference Signal (SRS). The positions of DM RS and SRS in the uplink resource blocks is illustrated in Fig. 2.17.

DM RS are located in symbols 3 and 10 of each subframe. When explicitly requested by the eNodeB, SRS signals of ZC sequences are sent in the symbol 13 location. The SRS for a given UE is located every 2 subcarriers using the Interleaved FDMA (IFDMA) method. A 10-bit PDCCH message describes the time, period (between 2 and 320 ms) and bandwidth for the UE to send the SRS message. For more details on SRS, see [30], p. 370. As both DM RS and SRS are Constant Amplitude Zero AutoCorrelation (CAZAC) sequences, two UEs can send each their SRS using the same resources provided they use different cyclic shifts, making sequences orthogonal (Sect. 2.4.3). The eNodeB is then able to separate the two different pilot signals. Figure 2.17 displays an example of three subframes with UE 1 transmitting data on one fixed PRB with a SRS cycle of more than 2 ms and UE 2 sending data on two fixed PRBs with a SRS cycle of 2 ms.

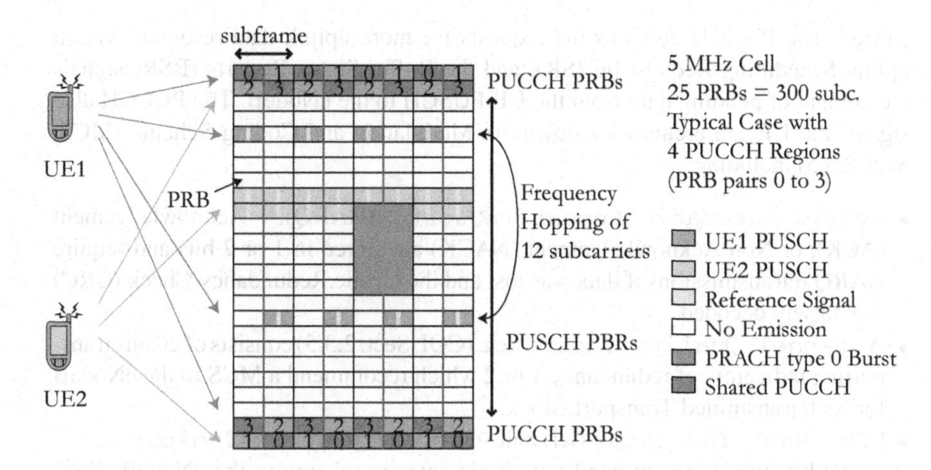

Fig. 2.18 Uplink channels

2.4.2 Uplink Physical Channels

Uplink data and control streams are sent using three physical channels allocated on the system bandwidth:

- The `Physical Uplink Shared Channel` (PUSCH) carries all the data received by a eNodeB from its UEs. It can also carry some control. LTE PUSCH only supports localized UE allocations, i.e. all PRBs allocated to a given UE are consecutive in frequency.
- The `Physical Uplink Control Channel` (PUCCH) carries the major part of the transmitted control values via uplink.
- `Physical Random Access Channel` (PRACH) carries the connection requests from unconnected UEs.

Figure 2.18 illustrates the localization of the physical channels for the example of a 5 MHz cell. The PUCCH is localized on the edges of the bandwidth while PUSCH occupies most of the remaining PRBs. PUCCH information with redundancy is sent on pairs of PRBs called regions (indexed in Fig. 2.18), where the two PRBs of a region are located on opposite edges. This technique is called frequency hopping and protects the PUCCH against localized frequency selectivity. There are typically 4 PUCCH regions per 5 MHz band. However, the eNodeB can allocate the PUSCH PRBs in PUCCH regions if they are not needed for control. Multiple UE control values can be multiplexed in PUCCH regions via Code Division Multiplexing (CDM).

Uplink decisions are generally taken by the eNodeB MAC scheduler and transmitted via downlink control channels. Consequently, the UEs are not required to send back these eNodeB decisions to the eNodeB, thus reducing the overhead of the PUCCH. 7 PUCCH region formats are defined according to the kind of information carried. The PUCCH contains the requests for more uplink data resources via an uplink Scheduling Request bit (SR), and the Buffer Status Reports (BSR) signals the amount of pending data from the UE PUSCH to the eNodeB. The PUCCH also signals the UE preferences for downlink Modulation and Coding Scheme (MCS, Sect. 2.3.5) including:

- `Hybrid Automatic Repeat reQuest` (HARQ), Acknowledgement (ACK), or Non Acknowledgement (NACK) are stored in 1 or 2 bits and require HARQ retransmissions if data was lost and the Cyclic Redundancy Check (CRC) was falsely decoded.
- A `Channel Quality Indicator` (CQI, Sect. 2.3.5) consists of 20 bits transporting indicators of redundancy 1 or 2 which recommend a MCS to the eNodeB for each transmitted Transport Block.
- `MIMO Rank Indicator` (RI) and `Precoding Matrix Indicator` (PMI) bits which recommend a multiple antenna scheme to the eNodeB (Sect. 2.3.6).

The PUCCH carries this control information periodically. If no PRB is available in the PUCCH channel, the PUSCH can carry some control information. Additionally, the eNodeB can request additional aperiodic CQI and link adaptation reports. These reports are sent in PUSCH and can be much more complete than periodic reports (up to 64 bits).

The PRACH is allocated periodically over 72 subcarriers (6 PRBs, Fig. 2.18). Employing 72 subcarriers is favorable because it has the widest bandwidth available for all LTE configurations between 1.4 MHz and 20 MHz. The PRACH burst can last between 1 and 3 subframes depending on the chosen mode; long modes are necessary for large cells. The PRACH period depends on the adjacent cell configurations and is typically several subframes. See Sect. 2.4.5 and [30], p. 421 for more details on the PRACH.

The uplink is shared between several transmitting UEs with different velocities and distances to the eNodeB. Certain precautions must be taken when multiple UEs are transmitting simultaneously, one of which is the `timing advance`. Timing advance consists of sending data with the corrected timing, so compensating for the propagation time and allowing the synchronization of all data received at the eNodeB. The correct timing advance is evaluated using the timing received from the PRACH bursts.

2.4.3 Uplink Reference Signals

In uplink communication, a Demodulation Reference Signal (DM RS) is constructed from a set of complex-valued Constant Amplitude Zero AutoCorrelation (CAZAC) codes known as Zadoff–Chu (ZC) sequences [17, 26]. ZC sequences are computed with the formula:

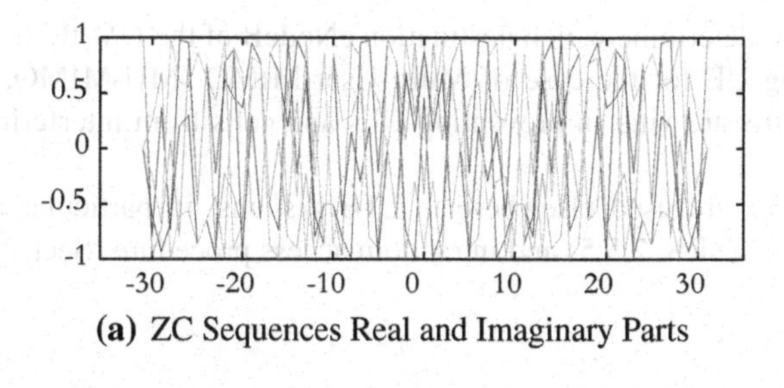

(a) ZC Sequences Real and Imaginary Parts

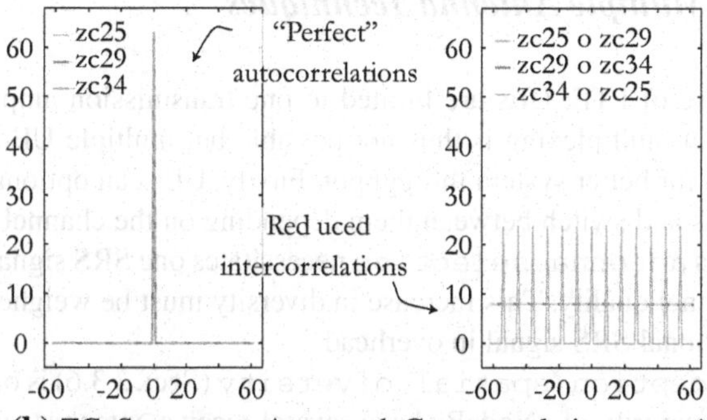

(b) ZC Sequences Auto and Crosscorrelation magnitudes

Fig. 2.19 Length-63 Zadoff–Chu sequences with index 25, 29 and 34

$$z_q(n) = \exp\left(\frac{-j\pi qn(n+1)}{N_{ZC}}\right) \tag{2.6}$$

where n is the sample number, q is the sequence index ranging from 1 to $N_{ZC} - 1$, and N_{ZC} is the sequence size. Three ZC sequences of length 63 with indexes 25, 29 and 34 are illustrated in Fig. 2.19. Their amplitude and the amplitude of their Fourier transform are constant, keeping low the PAPR of their transmission (Fig. 2.19a). All three sequences have an autocorrelation close to the Dirac function, enabling the creation of several orthogonal sequences from a single sequence using cyclic shifts. It may also be noted that the cross correlation between two ZC sequences with different indices is small compared to the autocorrelation peak (Fig. 2.19b). Consequently, two sequences can be decoded independently when sent simultaneously as long as their indices or cyclic shifts are different.

Each uplink DM RS of length N_P consists of a ZC sequence with the highest prime size smaller than N_P. The ZC sequence is cyclically extended to N_P elements and then cyclically shifted by α elements. For the smallest DM RS sizes of 12 and 24 elements, special codes replace ZC sequences because they outperform the ZC sequences in these cases ([30], p. 361). DM RS primarily serve to estimate the channel. They also carry information in their sequence index q and cyclic shift α, namely an ID to determine which destination eNodeB of the DM RS and to identify the transmitting UE for the case of Multi-User MIMO (MU-MIMO, Sect. 2.4.4). Indices and shifts are vital in preventing UEs and cells from interfering with each other.

ZC sequences are used elsewhere in LTE systems, in particular as Sounding Reference Signal (SRS, 2.5.5) and in random access procedure (Sect. 2.4.5).

2.4.4 Uplink Multiple Antenna Techniques

In the first release of LTE, UEs are limited to one transmission amplifier. Uplink single-user spatial multiplexing is thus not possible but multiple UE antennas can still be exploited for better system throughput. Firstly, UEs can optionally have two transmit antennas and switch between them depending on the channel quality. This method known as `antenna selection` necessitates one SRS signal per antenna to report on channel quality. This increase in diversity must be weighed against the cost of this additional SRS signal in overhead

Secondly, `reception spatial diversity` (Sect. 2.3.6) is often exploitable because the majority of eNodeBs have several receive antennas, each of which have known channel responses h_i. Naming the received values:

$$y = hx, \tag{2.7}$$

where each antenna stream has already been equalized (Sect. 2.3.2) and thus $h = [h_1 h_2 ... h_N]^T$ is the vector of channel responses, each being a scalar and $x = [x_1 x_2 ... x_N]^T$ and $y = [y_1 y_2 ... y_N]^T$ are the transmitted and received signal vectors across N antennas. The MRC detected signal is given by:

$$\hat{x} = \frac{h^H y}{h^H h} \tag{2.8}$$

where $\hat{x}$ is the complex detected symbol. MRC favors antennas which can receive high power signals.

Thirdly, `Multi-User MIMO` (MU-MIMO) also called Spatial Division Multiple Access (SDMA) consists of allocating the same resources to 2 UEs in the eNodeB and using the channel differences in frequency selectivity between UEs to separate the signals while decoding. Using MIMO can greatly increase the data rate and only requires special processing in the eNodeB. It necessitates orthogonal uplink DM RS reference sequences with different cyclic shift to independently evaluate the channel of each UE. Eight different DM RS cyclic shifts are defined in LTE for this purpose. This scheme is often called Virtual MU-MIMO because no added complexity is required at the UE: it is not necessary for the UE to know it shares the same resources with another UE. The complexity increases only at eNodeB side.

A 2×2 MU-MIMO scheme is equivalent to that shown in Fig. 2.14b but with the transmit antennas connected to separate UEs. Spatial multiplexing decoding, also known as MIMO detection, consists of reconstructing an estimate vector $\hat{x}$ of the sent signal vector x from y and H (Fig. 2.14b) in the eNodeB. The four channels in H can be considered to be flat fading (not frequency selective) because they have been individually equalized upon reception (Sect. 2.3.2); the channel is thus constant over all SC-FDMA subcarriers. The two most common low complexity linear MIMO detectors are that of `Zero Forcing` (ZF) and of `Minimum Mean-Square Error` (MMSE). In the ZF method, where $\hat{x} = Gy$ is the vector of detected symbols and y is the vector of received symbols, G is computed as the pseudo inverse of H:

$$G = (H^H H)^{-1} H^H. \tag{2.9}$$

The ZF method tends to amplify the transmission noise. The MMSE method is a solution to this problem, with:

$$G = (H^H H + \sigma^2 I)^{-1} H^H. \tag{2.10}$$

where σ^2 is the estimated noise power and I the identity matrix. Many advanced MIMO decoding techniques [23, 32] exist, notably including Maximum Likelihood Receiver (MLD) and Sphere Decoder (SD).

2.4.5 Random Access Procedure

The random access procedure is another feature of the uplink. While preceding features enabled high performance data transfers from connected UEs to eNodeBs, the random access procedure connects a UE to a eNodeB. It consists of message exchanges initiated by an uplink message in the `PRACH` channel (Sect. 2.4.2). It has two main purposes: synchronizing the UE to the base station and scheduling the UE for uplink transmission. The random access procedure enables a UE in idle mode to synchronize to a eNodeB and become connected. It also happens when a UE in connected mode needs to resynchronize or to perform a handover to a new eNodeB. The scheduling procedures uses the MME (Sect. 2.2.1) which is the network entity managing paging for phone calls. When a phone call to a given UE is required, the MME asks the eNodeBs to send paging messages with the UE identity in the PDCCH. The UE monitors the PDCCH regularly even in idle mode. When paging is detected, it starts a random access procedure. The random access procedure starts when the UE sends a PRACH signal.

Special time and frequency resources are reserved for the PRACH. Depending on the cell configuration, a PRACH can have a period from 1 to 20 ms. The signal has several requirements. One is that an eNodeB must be able to separate signals from several UEs transmitted in the same allocated time and frequency window. Another constraint is that the eNodeB must decode the signal rapidly enough to send back a Random Access Response (RAR) in PDSCH. A typical time between the end of PRACH reception and RAR is 4 ms ([30], p. 424). The PRACH message must also be well protected against noise, multipath fading and other interference in order to be detectable at cell edges. Finally, the resources dedicated to PRACH must induce a low overhead.

The chosen PRACH signals are ZC sequences of length 839, except in the special TDD format 4, which is not treated here. Good properties for ZC sequences are explained in Sect. 2.4.3 where their use in reference signals is described. A set of 64 sequences (called signatures) is attributed to each cell (out of 838 sequences in total). A signature is a couple (n_S, n_{CS}) where $n_S \leq N_S \leq 64$ is the root index and $n_{CS} \leq N_{CS} \leq 64$ is the cyclic shift so that $N_{CS} = \lceil 64/N_S \rceil$. A combination of a root and a cyclic shift gives a unique signature out of 64. A tradeoff between a high number of roots and a high number of cyclic shifts must be made. It will be seen that for more cyclic shifts, the easier the decoding becomes (Sect. 7.3). Each eNodeB broadcasts the first of its signatures and the other signatures are deduced by the UE from a static table in memory ([9], p. 39). A UE sends a PRACH message, by sending one of the 64 signatures included in the eNodeB signature set.

A PRACH message occupies 6 consecutive PRBs in the frequency domain, regardless of the cell bandwidth, during 1–3 subframes. This allocation is shown in Fig. 2.18. A UE sends PRACH messages without knowing the timing advance. A Cyclic Prefix (CP) large enough to cover the whole round trip must be used to protect the message from ISI with previous symbol. A slightly bigger Guard Time (GT) with no transmission must also be inserted after sending the PRACH message to avoid ISI with next symbol. GT must be greater than the sum of round trip and maximum delay spread to avoid overlap between a PRACH message and the subsequent signal. Depending on the cell size, the constraints on CP and GT sizes are different. Five modes of PRACH messages exist allowing adaption to the cell configuration ([11], p. 31). The smallest message consists of the 839 samples of one signature sent in one subframe over $800\,\mu s = 30720 * 0.8\,T_S = 24576\,T_S$ with CP and GP of about 100 μs. This configuration works for cells with a radius under 14 km and thus a round trip time under $14 * 2/300000 = 93\,\mu s$. The longest message consists of 1697 samples (twice the 839 samples of one signature) sent in 3 subframes over $1600\,\mu s = 30720 * 1.6\,T_S = 49152\,T_S$ with CP and GP of about 700 μs. This configuration is adapted to cells with a radius up to 100 km and thus a round trip time up to $100 * 2/300000 = 667\,\mu s$.

In 100 km cells, the round trip time can be almost as long as the transmitted signature. Thus no cyclic shift can be applied due to the ambiguity when receiving the same root with different cyclic shifts: the signal could result from separate signatures or from UEs with different round trips. The smaller the cell is, the more cyclic shifts can be used for one root resulting in a less complex decoding process and greater orthogonality between signatures (see auto and intercorrelations in Fig. 2.19b).

In the frequency domain, the six consecutive PRBs used for PRACH occupy $6 * 0.180 = 1.08$ MHz. The center band of this resource is used to send the 839 signature samples separated by 1.25 kHz. When the guard band is removed from the signal, it may be seen that the actual usable band is $839 * 1.25 = 1048.75$ kHz.

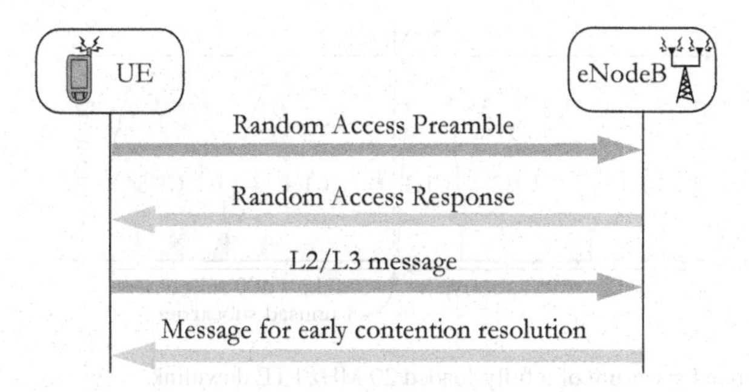

Fig. 2.20 Contention-based random access procedure

Two modes of random access procedure exist: the most common is the contention-based mode in which multiple UEs are permitted to send a PRACH signal in the same allocated time and frequency window. Contention-free mode is the other possibility and for this case, the eNodeB ensures that only a single UE can send a given ZC sequence for a given time and frequency window. The contention-free mode will not be presented here; see [30] for more details.

Figure 2.20 illustrates the contention-based random access procedure. Within the UE, the physical layer first receives a PRACH message request from upper layers with a PRACH starting frequency and mode, message power and parameters (root sequences, cyclic shifts...). A temporary unique identity called Random Access Radio Network Temporary Identifier (RA-RNTI) identifies the PRACH time and frequency window. Then, the UE sends a PRACH message using the given parameters. The UE monitors the PDCCH messages in subframes subsequent to the PRACH burst and if the RA-RNTI corresponding to that UE is detected, the corresponding PDSCH PRBs are decoded and RAR information is extracted. If no PDCCH message with the RA-RNTI of the UE is detected, the random access procedure has failed, and so a new PRACH message will be scheduled. More details on PRACH can be found in [11], p. 16.

An 11-bit timing advance is sent by the eNodeB to the UE within the RAR message ([11], p. 8). This timing advance is derived from the downlink reception of the previous downlink message, and ranges from 0 to 0.67 ms (the round trip in a 100 km cell) with a unit of 16 $T_s = 0.52\ \mu s$. When the position of the UE is changed, its timing advance alters. Timing advance updates can be requested by the eNodeB in MAC messages if a modification of the reference signal reception timing is measurable by the eNodeB. These update messages are embedded in PDSCH data.

In the RAR, the eNodeB allocates uplink resources and a Cell Radio Network Temporary Identifier (C-RTNI) to the UE. The first PUSCH resource granted to the UE is used to send L2/L3 messages carrying the random access data: these include connection and/or scheduling requests, and the UE identifier.

The final step in the contention-based random access procedure is the downlink contention resolution message in which the eNodeB sends the UE identifier corresponding to the successful PRACH connection. A UE which does not receive

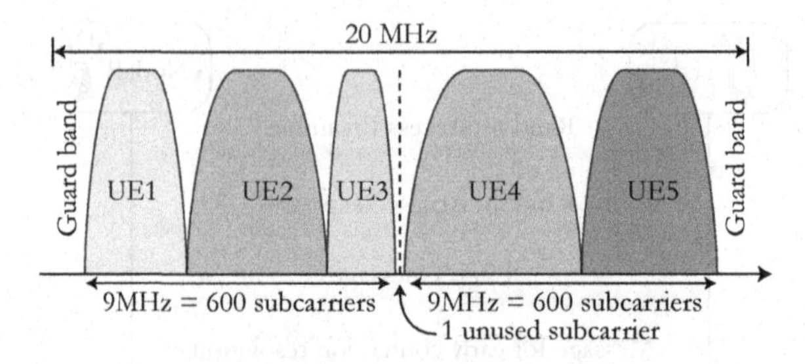

Fig. 2.21 Baseband spectrum of a fully-loaded 20 MHz LTE downlink

a message which includes its own identifier will conclude that the random access procedure has failed, and will restart the procedure.

2.5 LTE Downlink Features

2.5.1 Orthogonal Frequency Division Multiplexing Access

In contrast to uplink communications, a single entity, the eNodeB, handles the transmission over the whole bandwidth for the downlink. Transmission over a wide frequency band (up to 20MHz) and use of several antennas (4 or more) is possible because eNodeBs are powered by mains electricity and so the RF constraints are reduced compared to a UE. Consequently, a higher PAPR is allowed in the downlink than in the uplink ([30], p. 122).

Figure 2.21 displays the frequency use of a fully-loaded 20 MHz LTE downlink with localized UE allocation. A total bandwidth of 18 MHz is effectively used by the 1200 subcarriers, with the central subcarrier left unused because it may be "polluted" by RF oscillator leakage. The transmitted signal must fit within the transmission mask defined in [9]. Depending on bandwidth, the number of subcarriers varies according to Table 2.1. The number of PRBs per slot ranges from 6 in a 1.4 MHz cell to 100 in a 20 MHz cell. A configuration of 110 PRBs per slot in a 20 MHz cell with guard band reduction is also possible to increase data rates at the cost of a more complex radio frequency management. The number of subcarriers in the sampling band is a function of power of 2 (except for the 15 MHz case). A consequence of a subcarrier number power of two is that OFDMA and SC-FDMA baseband processing (Sect. 2.3.4) can be executed faster by employing the Fast Fourier Transform (FFT) algorithm to convert data into the Fourier domain.

Figure 2.22 shows that non-contiguous PRBs can be allocated to a UE using the downlink communication stream, enhancing protection against frequency selectivity at the cost of increasing control information. Frequency hopping between two PRBs in a single frame (for the case where two UEs exchange their PRBs) can also

Table 2.1 LTE downlink bandwidth configurations

Bandwidth(MHz)	1.4	3	5	10	15	20
Resource blocks per slot	6	15	25	50	75	100
Number of data subcarriers	72	180	300	600	900	1200
Used bandwidth (MHz)	1.08	2.25	4.5	9	13.5	18
Minimal sampling rate (MHz)	1.92	3.84	7.68	15.36	23.04	30.72
Number of subcarriers in sampling band	128	256	512	1024	1536	2048

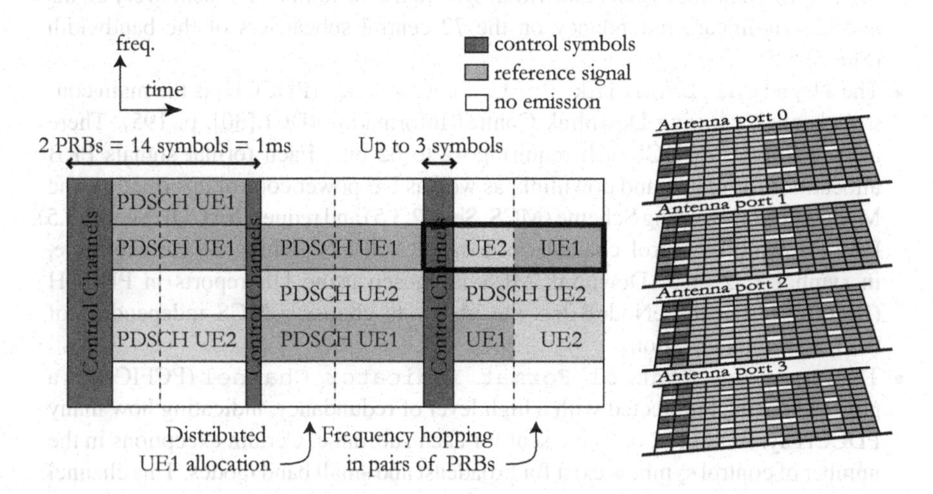

Fig. 2.22 Downlink multiple user scheduling and reference signals

reinforce this protection [4]. Note that all PRBs associated with a single UE have identical modulation and coding schemes because little gain increase is seen if PRBs of one UE have different modulation and coding schemes [2]. Figure 2.22 also shows the downlink cell specific reference signals that are inserted in symbols 0, 1 and 4 of each slot. Antenna ports 1 and 2 insert 4 reference values each per slot. Antenna ports 3 and 4 insert only 2 reference values each per slot. For each additional antenna, a reference signal overhead is added but multiple antennas can bring significant gain compensating for this throughput reduction (Sect. 2.3.6). The maximum possible reference signal overhead is: $(2*4+2*2)/(7*12) = 14.3\,\%$. Reference signals must be interpolated to be used in coherent decoding; they must consequently reflect most of the channel properties by covering the entire time/frequency resources. Using the channel coherence bandwidth and channel coherence time worst case estimation, the diamond shape localization of reference signals was chosen as a tradeoff between overhead and channel estimation accuracy. Downlink RS are length-31 Gold sequences (Sect. 2.5.3) initialized with the transmitting cell identifier. An eNodeB can send additional UE-specific RS (for beam forming) or broadcast RS (for broadcasted data) ([30], p. 163).

2.5.2 Downlink Physical Channels

Like in uplink communications, the downlink bits are transmitted through several physical channels allocated to specific physical resources:

- The `Physical Broadcast Channel` (PBCH) broadcasts basic information about the cell. It is a low data rate channel containing the Master Information Block (MIB), which includes cell bandwidth, system frame number. It is sent every 10 ms and has significant redundancy on the 72 central subcarriers of the bandwidth (Sect. 2.5.5).
- The `Physical Downlink Control Channel` (PDCCH) is the main control channel, carrying Downlink Control Information (DCI [30], p. 195). There are ten formats of DCI each requiring 42 to 62 bits. Each format signals PRB allocations for uplink and downlink, as well as UE power control information, the Modulation and Coding Scheme (MCS, Sect. 2.3.5) and request for CQI (Sect. 2.3.5). Every downlink control channels is located at the beginning of the subframe, in symbol 1, 2 or 3. Downlink MCS is chosen using UE reports in PUCCH (Sect. 2.4.1) but the eNodeB has the liberty to choose a MCS independent of the UE recommendation.
- The `Physical Control Format Indicator Channel` (PCFICH) is a physical channel protected with a high level of redundancy, indicating how many PDCCH symbols (1, 2 or 3) are sent for each subframe. Certain exceptions in the number of control symbols exist for broadcast and small band modes. This channel is transmitted in symbol 0, "stealing" resources from PDCCH.
- The `Physical Downlink Shared Channel` (PDSCH) is the only data channel, which carries all user data. There are seven PDSCH transmission modes, which are used depending multiple antenna usage, as decided by the eNodeB (Sect. 2.5.4). The transmission mode is part of the DCI sent in PDCCH. PDSCH also infrequently carries System Information Blocks (SIB) to complete the MIB information of the cell in PBCH channel.
- The `Physical Hybrid ARQ Indicator Channel` (PHICH) carries uplink HARQ ACK/NACK information, which request retransmission of uplink data when the Cyclic Redundancy Check (CRC) was incorrectly decoded (Sect. 2.3.5.1). CRC is added to an information block to detect infrequent errors, producing a small amount of redundancy. This process is called Forward Error Correction (FEC, Sect. 2.3.5). PHICH is sent using the same symbols as PDCCH.
- The `Physical Multicast Channel` (PMCH) is used to broadcast data to all UEs via Multimedia Broadcast and Multicast Services (MBMS). This special mode was especially created to broadcast television, and is not considered in this document.

A UE must constantly monitor the control channels (PDCCH, PCFICH, PHICH). Due to the compact localization of control channels in the first few symbols of each subframe, a UE can execute some "micro sleeps" between two control zones, and thus save power when no PRB with the identifier of the UE was detected in

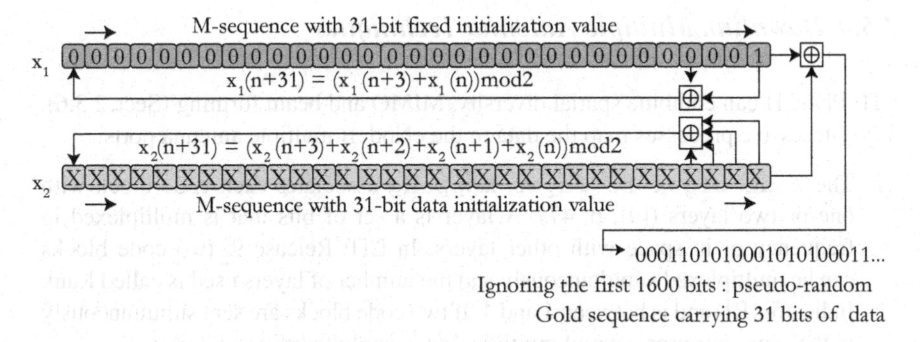

Fig. 2.23 Gold pseudo random sequence generation

the subframe. Special modulation and coding schemes are used for control channels. PDCCH is scrambled, and adds bitwise length-31 Gold sequences, equivalent to those used in downlink reference signals, initialized with UE identifier. Gold sequences are specific bit sequences that are explained in the next section. Several MU-MIMO UEs can receive PDCCH on the same PRBs using Code Division Multiplexing (CDM). PBCH and PDCCH are coded with convolutional code instead of turbo code because convolutional codes are better suited for small blocks (Sect. 2.3.5 and [30], p. 237).

2.5.3 Downlink Reference Signals

In downlink communication, three types of reference signals are sent: cell specific, UE specific and broadcast specific RS (2.5.1). Only the cell specific RS, which are the most common are considered in this study. Downlink RS are constructed from length-31 Gold sequences. `Gold sequences` are generated using 31-bit shift registers and exclusive ORs (Fig. 2.23). x_1 and x_2 are called maximum length sequences or M-sequences, and are spectrally flat pseudo-random codes but can carry up to 31 bits of data. While x_1 initial value is a constant, x_2 initial value carries the data. In order for two Gold codes with close initialization to be orthogonal, the 1600 first outputs are ignored. The resulting signal shift improves the code PAPR and the two Gold codes from different eNodeBs are orthogonal, enabling a UE to decode the signal reliably [5]. The PAPR of reference signals is very important because the reference signals are often power boosted compared to data.

The initialization value of a Gold sequence is generated from the physical layer cell identity number, the slot number within the radio frame, the symbol number within the slot, and an indicator of normal or extended CP. The Gold sequence bits are QPSK modulated and transmitted on RS allocated resources. The same length-31 Gold bit sequences are used for scrambling data over the frequency band in Sect. 2.5.1.

2.5.4 Downlink Multiple Antenna Techniques

LTE PDSCH can combine spatial diversity, MIMO and beam forming (Sect. 2.3.6). Two successive processes map the data to the eNodeB multiple antenna ports:

1. The `code block to layer mapping` associates each code block with one or two layers ([9], p. 47). A layer is a set of bits that is multiplexed in frequency or in space with other layers. In LTE Release 9, two code blocks can be multiplexed simultaneously and the number of layers used is called Rank Indicator (RI) and is between 1 and 4. If two code blocks are sent simultaneously in the same resource, spatial multiplexing is exploited (Sect. 2.3.6).
2. The `precoding` jointly processes each element of the layers to generate the n_T antenna port signals. It consists of multiplying the vector containing one element from each of the RI layers with a complex precoding matrix W. An antenna port can be connected to several antennas but these antennas will send an identical signal and so be equivalent to a single antenna with improved frequency selectivity due to diversity. The number of antenna ports n_T is 1, 2 or 4. In LTE, precoding matrices are chosen in a set of predefined matrices called a "codebook" ([9], p. 48). Each matrix has an index call Precoding Matrix Indicator (PMI) in the current codebook.

Figure 2.24 illustrates the layer mapping and precoding for `spatial diversity` for different numbers of RI and n_T. `Space-Frequency Block Coding` (SFBC) is used to introduce redundancy between several subcarriers. SFBC is simply introduced by temporally separating the values such as Space-Time Block Coding (STBC, Sect. 2.3.6) prior to the IFFT which is executed during OFDMA encoding. In Fig. 2.24a, one Code Block is sent to two antennas via two layers. The precoding matrix used to introduce SFBC follows the Alamouti scheme [15]. In Fig. 2.24b, a derived version of the Alamouti scheme is adapted for transmission diversity with SFBC over four antenna ports. Of every four values, two are transmitted using a pair of antenna ports with certain subcarriers and the two remaining values are transmitted on the other pair of antennas over different subcarriers. Antenna ports are not treated identically: antenna ports 3 and 4 transmit less reference signal so due to the resulting poorer channel estimation, these ports must not be paired together.

There are 7 PDSCH modes defining how the eNodeB exploits multiple antennas in its communication with a UE:

1. In `transmission mode 1`, no multiple antenna technique is used.
2. In `transmission mode 2`, there are several antennas but no spatial multiplexing. One Transport Block is sent per TTI with antenna diversity (Fig. 2.24a and b).
3. In `transmission mode 3`, open loop spatial multiplexing (MIMO) is used, i.e. the UE does not feed back information that would enable UE-specific precoding (i.e. beam forming). This precoding scheme is known as Cyclic Delay Diversity (CDD). It is equivalent to sending each subcarrier in a different direction, increasing frequency diversity of each transmitted Code Block.

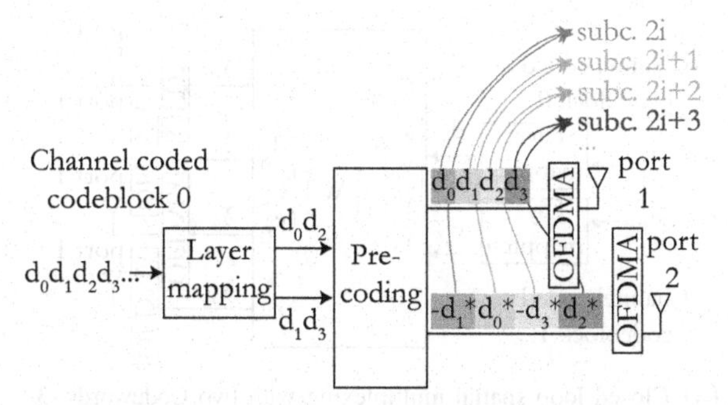

(a) SFBC with one Codeword, 2 Layers and 2 Antennas

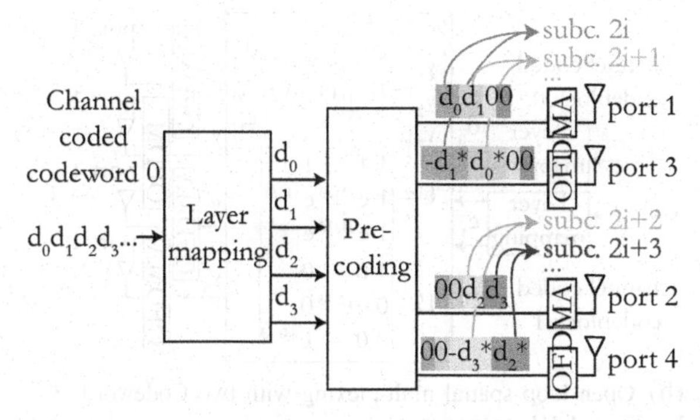

(b) SFBC with one Codeword, 4 Layers and 4 Antennas

Fig. 2.24 Layer mapping and precoding for spatial diversity with different multi-antenna parameters

4. In transmission mode 4, closed loop spatial multiplexing (MIMO) is used. The UE feeds back a RI and a PMI to advise the eNodeB to form a beam in its direction, i.e. by using the appropriate transmit antennas configuration. It also reports one Channel Quality Indicator (CQI) per rank, allowing the eNodeB to choose the MCS for each Transport Block.
5. In transmission mode 5, MU-MIMO is used (Sect. 2.4.4). Each of two UEs receives one of two transmitted Transport Blocks in a given resource with different precoding matrixes.
6. In transmission mode 6, there is no spatial multiplexing (rank 1) but UE precoding feedback is used for beamforming.
7. Transmission mode 7 corresponds to a more accurate beam forming using UE-specific downlink reference signals.

Examples of closed loop and open loop `spatial multiplexing` encoding are illustrated in Fig. 2.25. In Fig. 2.25a, the closed loop is used: the UE reports a PMI to assist in the eNodeB choice of a good precoding matrix W. In Fig. 2.25b, there is no UE feedback, so the two multiplexed code blocks are transmitted with a maximum

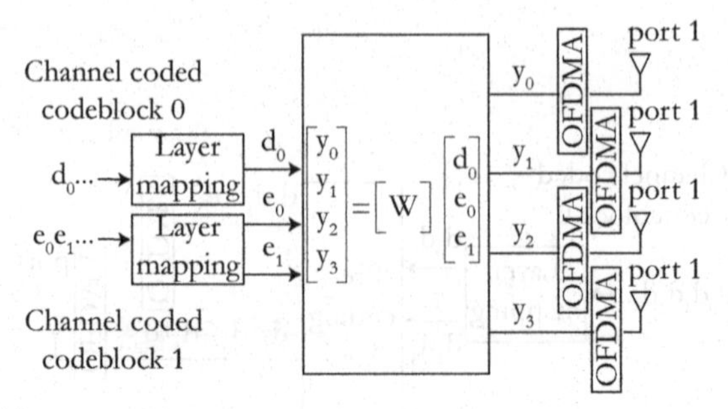

(a) Closed loop spatial multiplexing with two Codewords, 3 Layers, 4 Antennas

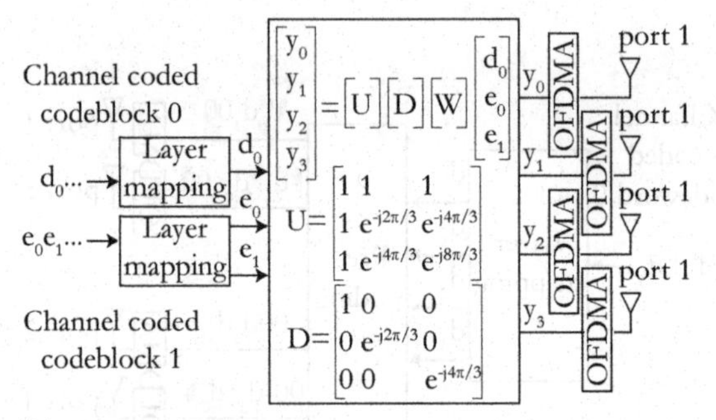

(b) Open loop spatial multiplexing with two Codeword, 3 Layers and 4 Antennas

Fig. 2.25 Layer mapping and precoding for spatial multiplexing with different multiantenna parameters

transmission diversity configuration with CDD. While matrix U mixes the layers, matrix D creates different spatial beams for each subcarrier which is equivalent to sending each subcarrier in a different directions. Finally, W is the identity matrix for two antenna ports and a table of predefined matrices used successively for four antenna ports. The objective of W is to decorrelate more the spatial layers.

A UE knows the current transmission mode from the associated PDCCH format. MIMO detectors for the LTE downlink are located in the UEs; the ZF or MMSE techniques presented in Sect. 2.4.4 are used for the decoding. This topic is not within the scope of this study. More information on LTE downlink MIMO detectors can be found in [33].

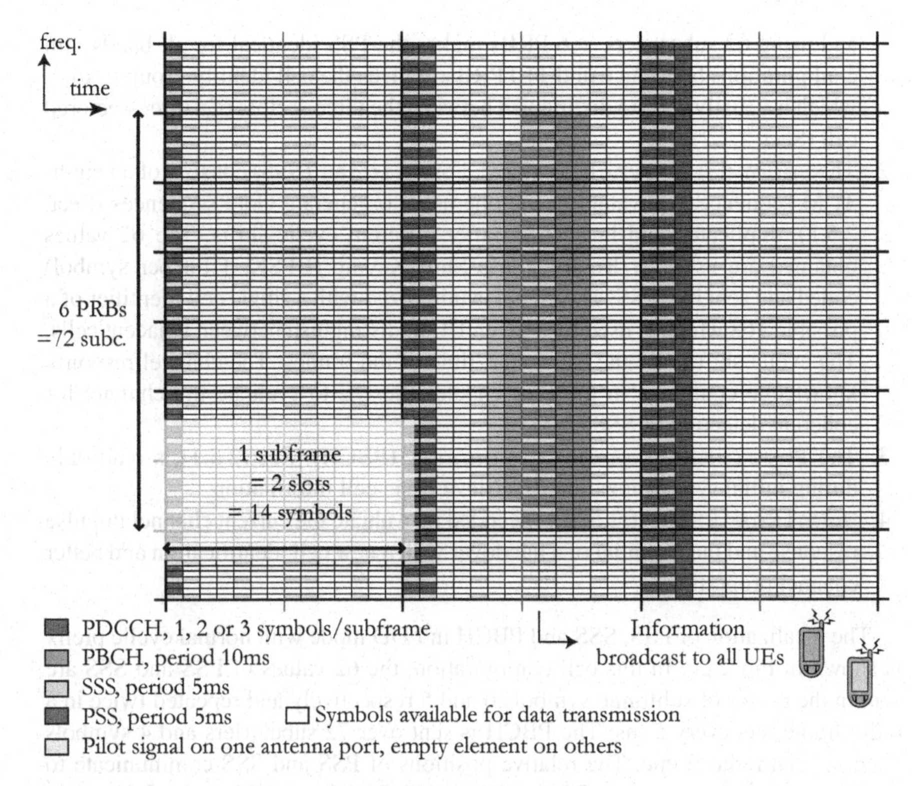

Fig. 2.26 Downlink PSS, SSS, RS and PBCH localization in subframes

2.5.5 UE Synchronization

Synchronizing the UE with the eNodeB is a prerequisite to data exchanges. A UE will monitor the eNodeB downlink synchronization signals when it requires a new connection (initial synchronization) or during a handover (new cell identification). The synchronization signals communicate the time basis as well as the cell identifier, bandwidth, cyclic prefix length and cell duplex mode. Four types of downlink signals must be decoded by the UE for synchronization: the Primary Synchronization Signal (PSS), the Secondary Synchronization Signal (SSS), the Physical Broadcast Channel (PBCH) and the downlink Reference Signal (RS). These signals give a UE information about the cell identity and parameters, as well as the quality of the channel.

1. The `Primary Synchronization Signal` (PSS) consists of a Zadoff–Chu (ZC) sequence with length N_{ZC} 63 and index q 25, 29 or 34. These sequences are illustrated in Fig. 2.19. The middle value is "punctured" to avoid using the DC subcarrier and so is sent twice in each frame using the 62 central subcarriers (Fig. 2.26). A ZC length of 63 enables fast detection using a 64 point FFT

and using 62 subcarriers ≤ 6 PRBs makes the PSS identical for all bandwidth configurations between 6 and 110 PRBs. q gives the cell identity group (0 to 2) which is usually used to distinguish between the three sectors of a three-sectored cell (Sect. 2.1.2).

2. The `Secondary Synchronization Signal` (SSS) consists of a length-31 M-sequence, equivalent to the one used to generate Gold sequences (Sect. 2.5.3), duplicated and interleaved after different cyclic shifts. The 62 values obtained are used for Binary Phase Shift Keying (BPSK, 1 bit per symbol) coded and sent like PSS. The signal is initialized with a single cell identifier of a possible 168 which is used by a given UE to distinguish between adjacent cells. The cyclic shift gives the exact start time of the frames. The channel response knowledge contained in the PSS enables the UE to evaluate the channel for coherent detection of SSS.
3. The `Physical Broadcast Channel` (PBCH) (Sect. 2.5.1) is read only during initial synchronization to retrieve basic cell information.
4. `Downlink RS` (Sect. 2.5.3) is decoded to evaluate the current channel impulse response and the potential of a handover with a new cell identification and better channel properties.

The localization of PSS, SSS and PBCH in FDD mode with normal cyclic prefix is shown in Fig. 2.26. In this cell configuration, the 62 values of PSS and SSS are sent in the center of subframe symbols 6 and 5 respectively and repeated twice in a radio frame, i.e. every 5 ms. The PBCH is sent over 72 subcarriers and 4 symbols once in each radio frame. The relative positions of PSS and SSS communicate to UEs the type of duplex mode (TDD, FDD or HD-FDD) and CP length of the cell.

This section detailed the features of the LTE physical layer. Next section introduces the concept of dataflow Model of Computation and details the models that are used in this study to represent the LTE eNodeB physical layer processing.

References

1. R1-050587 (2005) OFDM radio parameter set in evolved UTRA downlink
2. R1-060039 (2006) adaptive modulation and channel coding rate control for single-antenna transmission in frequency domain scheduling
3. R1-062058 (2006) E-UTRA TTI size and number of TTIs
4. R1-073687 (2007) RB-level distributed transmission method for shared data channel in E-UTRA downlink
5. R1-081248 (2008) PRS sequence generation for downlink reference signal
6. GSM-UMTS network migration to LTE (2010) Technical report, 3G, Americas
7. 36.101, G.T. (2009) Evolved universal terrestrial radio access (E-UTRA); user equipment (ue) radio transmission and reception (Release 9)
8. 36.104, G.T. (2009) Evolved universal terrestrial radio access (E-UTRA); base station (bs) radio transmission and reception (Release 9)
9. 36.211, G.T. (2009) Evolved universal terrestrial radio access (E-UTRA); physical channels and modulation (Release 9)

10. 36.212, G.T. (2009) Evolved universal terrestrial radio access (E-UTRA); multiplexing and channel coding (Release 9)
11. 36.213, G.T. (2009) Evolved universal terrestrial radio access (E-UTRA); physical layer procedures (Release 9)
12. 36.321, G.T. (2009) Evolved universal terrestrial radio access (E-UTRA); medium access control (mac) protocol specification (Release 9)
13. 36.322, G.T. (2009) Evolved universal terrestrial radio access (E-UTRA); radio link control (rlc) protocol specification (Release 9)
14. 36.323, G.T. (2009) Evolved universal terrestrial radio access (E-UTRA); packet data convergence protocol (pdcp) specification (Release 9)
15. Alamouti SM (2007) A simple transmit diversity technique for wireless communications. The best of the best: fifty years of communications and networking research, p 17
16. Berrou C, Glavieux A (2007) Near optimum error correcting coding and decoding: turbo-codes. The best of the best: fifty years of communications and networking research, p 45
17. Chu D (1972) Polyphase codes with good periodic correlation properties. IEEE Trans Inf Theory 18(4):531–532
18. Ciochina C, Mottier D, Sari H (2006) Multiple access techniques for the uplink in future wireless communications systems. Third COST 289 Workshop
19. Cox R, Sundberg C (1994) An efficient adaptive circular viterbi algorithm for decoding generalized tailbiting convolutional codes. IEEE Trans Veh Technol 43(1):57–68. doi:10.1109/25.282266
20. Dahlman E, Parkvall S, Skold J, Beming P (2007) 3G evolution: HSPA and LTE for mobile broadband. Academic Press, Oxford
21. Holma H, Toskala A (2009) LTE for UMTS: OFDMA and SC-FDMA based radio access. Wiley, Chichester
22. Holter B (2001) On the capacity of the mimo channel-a tutorial introduction. In: IEEE Norwegian symposium on, signal processing, pp 167–172
23. Larsson EG (2009) MIMO detection methods: how they work. IEEE Signal Process Mag 26(3):9195
24. Mehlführer C, Wrulich M, Ikuno JC, Bosanska D, Rupp M (2009) Simulating the long term evolution physical layer. In: Proceedings of the 17th European signal processing conference (EUSIPCO 2009), Glasgow
25. Norman T (2009) The road to LTE for GSM and UMTS operators. Technical report, Analysys Mason
26. Popovic B (1992) Generalized chirp-like polyphase sequences with optimum correlation properties. IEEE Trans Inf Theory 38(4):1406–1409. doi:10.1109/18.144727
27. Pyndiah R (1997) Iterative decoding of product codes: block turbo codes. In: Proceedings of the 1st international symposium on turbo codes and related topics, pp 71–79
28. Rihawi B, Louet Y (2006) Peak-to-average power ratio analysis in MIMO systems. Information and communication technologies, 2006, ICTTA'06, vol 2
29. Rumney M (2009) LTE and the evolution to 4G wireless: design and measurement challenges. Wiley, New York
30. Sesia S, Toufik I, Baker M (2009) LTE, the UMTS long term evolution: from theory to practice. Wiley, Chichester
31. Shannon CE (2001) A mathematical theory of communication. ACM SIGMOBILE Mobile Comput Commun Rev 5(1):55
32. Trepkowski R (2004) Channel estimation strategies for coded MIMO systems, M.S. thesis. Ph.D. thesis, Virginia Polytechnic University, Blacksburg, Va., June 2004
33. Wu D, Eilert J, Liu D (2009) Evaluation of MIMO symbol detectors for 3GPP LTE terminals. In: Proceedings of 17th European signal processing conference (EUSIPCO), Glasgow

Chapter 3
Dataflow Model of Computation

3.1 Introduction

To study the LTE physical layer on multi-core architectures, a Model of Computation (MoC) is needed to specify the LTE algorithms. This MoC must have the necessary expressivity, must show the algorithm parallelism and must be capable of locating system bottlenecks. Additionally, it must be simple to use and sufficiently intuitive for designers to manage rapid prototyping. In the following sections, the concept of Model of Computation is introduced and the models which will be used to describe LTE are detailed. After an overview, Sect. 3.2 focuses on the Synchronous Dataflow model (SDF), which forms the foundation of numerous MoCs as it can be precisely analyzed. Section 3.4 explains an extension of SDF which compactly expresses the behavior of an algorithm. Finally, Sect. 3.5 describes hierarchical dataflow models which enable compositions of sub-parts.

3.1.1 Model of Computation Overview

A Model of Computation (MoC) defines the semantics of a computational system model, i.e. which components the model can contain, how they can be interconnected, and how they interact. Every programming language has at least one (often several) underlying MoCs. A MoC describes a method to specify, simulate and/or execute algorithms. MoCs were much promoted by the Ptolemy and Ptolemy II projects from the University of California Berkeley. In [1], Chang, et al. explain how several MoCs can be combined in the Ptolemy tool. MoCs can serve three different purposes:

1. `Specification`: a specification model focuses on clearly expressing the functionalities of a system. It is especially useful in standards documents.
2. `Simulation`: a simulation model is used to extract knowledge of a system when the current implementation is not available. It may be much simpler than the final code and is focused precisely on the features of interest.

M. Pelcat et al., *Physical Layer Multi-Core Prototyping*,
Lecture Notes in Electrical Engineering 171, DOI: 10.1007/978-1-4471-4210-2_3,

3. `Execution`: an execution model must contain all the information for the final code execution.

These objectives are not mutually exclusive but it is sufficiently complex to obtain all three simultaneously. The International Technology Roadmap for Semiconductors (ITRS) predicts that unified models covering the three problems will be available around 2020 [2]. The unified model is called "executable specification".

The definition of MoCs is broad and covers many models that have emerged in the last few decades. The notion of a Model of Computation is close to the notion of a programming paradigm in the computer programming and compilation world. Arguably, the most successful MoC families, in terms of adoption in academic and industry worlds are:

- `Finite State Machine MoCs` (FSM) in which states are defined in addition to rules for transitioning between two states. FSMs can be synchronous or asynchronous [3]. The rules depend on control events and a FSM MoC is often used to model control processes. The actual computation modeled by a FSM is performed at transitions. The FSM MoC is often preferred for control-oriented applications. The imperative programming paradigm is equivalent to a FSM in which a program has a global state which is modified sequentially by modules. The imperative programming paradigm is the basis of the most popular programming languages including C. Its semantics are directly related to the Turing machine / Von Neumann hardware it targets. It is often combined with a higher-level Object-Oriented Programming (OOP) MoC, in particular in C++, Java and Fortran programming languages.
- `Process Network MoCs` (PN) in which concurrent and independent modules known as processes communicate ordered tokens (data quanta) through First-In First-Out (FIFO) channels. Process Network MoCs are often preferred to model signal processing algorithms. The notion of time is usually not taken into account in Process Network MoCs where only the notion of causality (who precedes whom) is important. The `dataflow Process Networks` that will be used in the rest of the document are a subset of Process Networks.
- `Discrete EventMoCs` (DE) in which modules react to events by producing events. Modules themselves are usually specified with an imperative MoC. These events are all tagged in time, i.e. the time at which events are consumed and produced is essential and is used to model system behavior. Discrete Event MoCs are usually preferred to model clocked hardware systems and simulate the behavior of VHDL and Verilog coded FPGA or ASIC implementations.
- `Functional MoCs` in which a program does not have a preset initial state but uses the evaluation result of composed mathematical functions. Examples of programming languages using Functional MoCs include Haskell, Caml and XSLT. The origin of functional MoCs lies in the lambda calculus introduces by Church in the 1930s. This theory reduces every computation to a composition of functions with only one input.
- `Petri Nets` which contain unordered channels named transitions, with multiple writers and readers and local states called places, storing data tokens. Transitions

and states are linked by directed edges. A transition of a Petri Net fires, executing computation and producing tokens on its output arcs, when a token is available at the end of all its input arcs [3, 4]. Comparing with a FSM MoC, the advantage of Petri Nets is their ability to express parallelism in the system while a FSM is sequential.

- `Synchronous MoCs` in which, like in Discrete Events, modules react to events by producing new events but contrary to Discrete Events, time is not explicit and only the simultaneity of events and causality are important. Programming languages based on Synchronous MoCs include Signal, Lustre and Esterel.

The MoCs that are considered in this document are all within the above `Process Network MoCs`. The previous list is not exhaustive. A more complete description of most of these MoCs can be found in [3].

3.1.2 Dataflow Model of Computation Overview

In this document, the physical layer of LTE, which is a signal processing application, is modeled. The targeted processing hardware is composed of Digital Signal Processors (DSP) and the signal arrives at this hardware after a conversion into complex numbers (Sect. 2.1.3). The obvious choice of MoC to model the processing of such a flow of complex numbered data is `Dataflow Process Network`. In [5], Lee and Parks define Dataflow Process Networks and locate them precisely in the broader set of Process Network MoCs. The MoCs discussed in following Sections are illustrated in Fig. 3.1. Each MoC is a tradeoff between expressivity and predictability. MoCs are divided into main branch models and branched off models:

- `Main branch models` as shown in Fig. 3.1 are all subsets of the Kahn Process Network (KPN) MoC [6], which adds constraints to the models that contain them. KPN consists of continuous processes that communicate through infinite lossless FIFO channels. The continuity property is explained in [5]. The outputs of a network of continuous processes can be computed iteratively, i.e. it can be factorized to one minimum "fixed point" [5], where the computation is repeated. An important property of KPN is that FIFO channel writes are non-blocking and FIFO channel reads are blocking. KPN is the only non-dataflow MoC in Fig. 3.1. Dataflow Process Networks additionally specify the behavior of their processes when they receive and send data. These processes are then called actors and fire when they receive data tokens, themselves producing data tokens. A set of firing rules defines when an actor fires. Firing and actor consists in starting its preemption-free execution. Other main branch models are restrictions of the KPN model with different firing rule definitions.
- `Branched off models` are derived from main branch models with added features adapting their targeted usage to a particular problem, meaning that these models are no longer subsets of higher-level main branch MoCs.

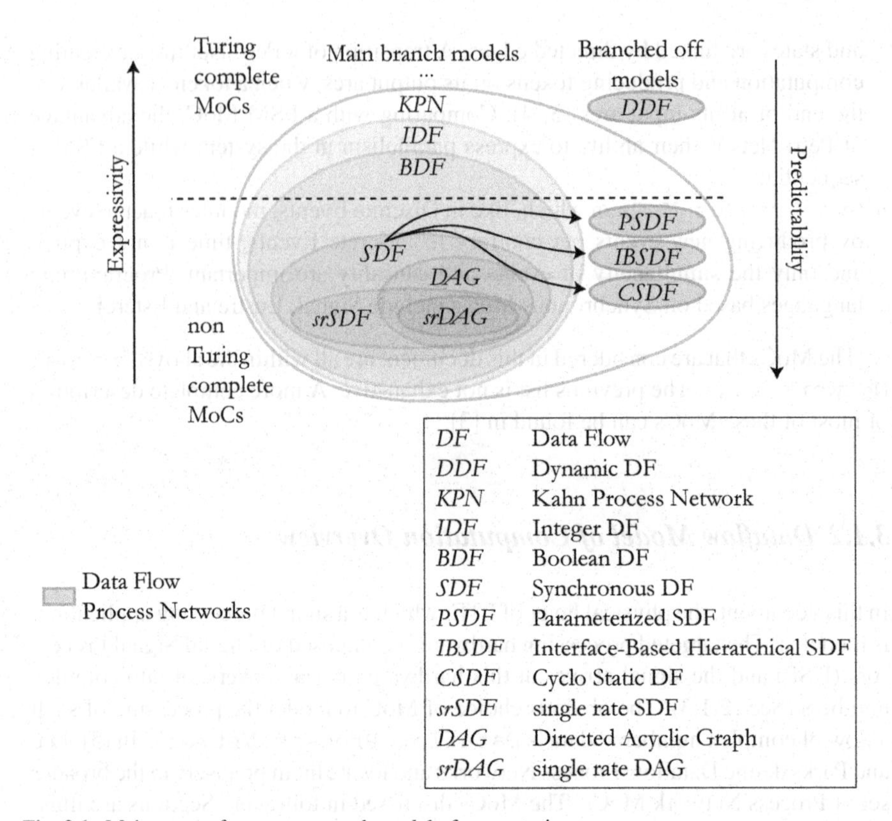

Fig. 3.1 Main types of process network model of computation

When a given MoC is Turing complete, this MoC expresses all possible algorithms. Figure 3.1 distinguishes Turing complete MoCs from those that are non-Turing complete. The non-Turing complete MoCs have limited expressiveness but they enable more accurate compile-time studies. Many Dataflow Process Network MoCs have been introduced in the literature, each offering a tradeoff between compile-time predictability and capacity to model dynamic run-time variability. They differ mostly by their firing rules:

- `Dynamic Data Flow` MoC (DDF) is a MoC with special firing rules and is thus a dataflow process network MoC; but a DDF actor can "peek" at a token without consuming it, thus defining non blocking reads not allowed in KPN. This property can make DDF non deterministic and complicate its study [5]. Compared with KPN, DDF has advantages as a dataflow model. Contrary to the KPN processes, DDF actors do not necessitate a multi-tasking OS to be executed in parallel. A simple scheduler that decides which actor to execute depending on the available tokens is sufficient. The CAL language software execution method is based on this property [7].

- `The Synchronous Dataflow` MoC (SDF) has been the most widely used dataflow MoC because of its simplicity and predictability. Each SDF actor consumes and produces a fixed number of tokens at each firing. However, it is not Turing complete and cannot express conditions in an algorithm. Figure 3.2a shows a SDF algorithm. The small dots represent tokens already present on the edges when the computation starts.
- `The single rate SDF` MoC is a subset of SDF. A single rate SDF graph is a SDF graph where the production and consumption of tokens on an edge are always equal (Fig. 3.2b).
- `The Directed Acyclic Graph` MoC (DAG) is also a subset of SDF. A DAG is a SDF graph where no path can contain an actor more than once, i.e. a graph that contains no cycle (Fig. 3.2c).
- `The single rate Directed Acyclic Graph` MoC is a DAG where productions and consumptions are equal.
- `The Boolean Dataflow` MoC (BDF) is a SDF with additional special actors called `switch` and `select` (Fig. 3.2d). The `switch` actor is equivalent to a demultiplexer and the `select` actor to a multiplexer. The data tokens on the unique input of a switch are produced on its true (respectively false) output if a control token evaluated to true (respectively false) is sent to its control input. The BDF model adds control flow to the SDF dataflow and it is this control that makes the model Turing complete.
- `The Integer Data Flow` MoC (IDF) is the equivalent of BDF but where the control tokens are integer tokens rather than Boolean tokens.
- `The Cyclo Static Dataflow` MoC (CSDF) extends SDF by defining fixed patterns of production and consumption (Fig. 3.2e). For instance, a production pattern of (1, 2) means that the actor produces alternatively one and two tokens when it fires.
- `The Parameterized SDF` MoC (PSDF) is a hierarchical extension of SDF that defines production and consumption relative to parameters that are reevaluated at reconfiguration points. Parameterized dataflow is a meta-modeling technique adding reconfigurability to a static model.
- `The Interface-Based Hierarchical SDF` MoC (IBSDF) is also a hierarchical extension of SDF that insulates the behavior of its constituent graphs, making them easy to combine and study. Interface-Based hierarchical dataflow is a meta-modeling technique which adds a hierarchy to a static model.

The above list is not complete. Many more dataflow MoCs have been defined and new models are studied in this document in the context of the PREESM tool development. Dataflow process networks only specify how actors communicate but cannot give the complete execution semantics (for example, an actor performing an addition still needs a description of this behavior to execute). If a Dataflow MoC is used for execution, it appears as a `coordination language` and must be combined with a `host language` defining actor behavior [5]. C code can be used as a software host language or Hardware Description Language (HDL) code as a hardware host language. CAL is a host language that can target both software

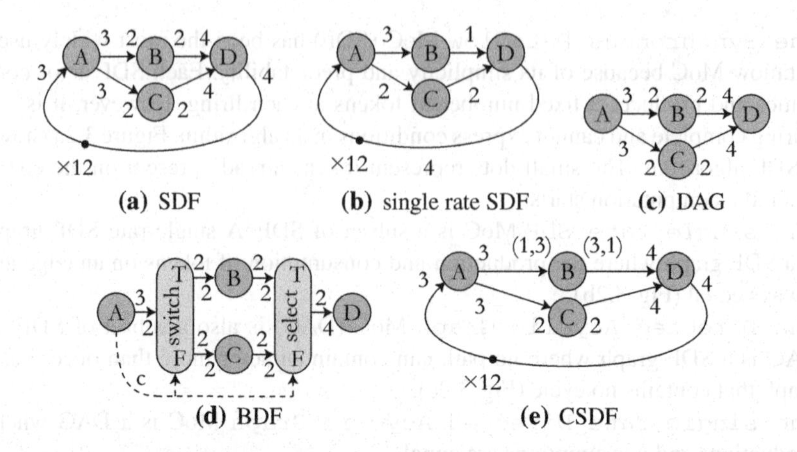

Fig. 3.2 A few dataflow MoC graph examples

and hardware implementation generation. In this study of LTE, C code is used as the host language but the code generation is flexible and may be adapted to other programming languages.

Process networks are usually employed to model one execution of a repetitive process. For a single application, descriptions with different `scales` are often possible. When modeling an H.264 60 Hz full-HD video decoder for instance, a graph describing the decoding of one 16×16 pixels macro-block can be created and executed 486,000 times per second. It is also possible to create a graph describing the decoding of one picture and repeat this description 60 times/s; or a graph describing the decoding of one group of pictures and which is then repeated several times per second. The right scale must be chosen, a scale which is too large produces graphs that are too large in terms of number of actors and complicates the study, whereas a scale that is too small will usually lead to sub-optimal implementations [8].

Aside from the scale, the `granularity` is another important parameter in describing an application. In [9], Sinnen defines the granularity G of a system as the minimum execution time of an actor divided by the maximum transfer time of an edge. Using this definition which depends on both algorithm and architecture, in a coarse grain system where $G \gg 1$, actors are seen to be "costly" compared to the data flowing between them. High granularity is a desirable property for system descriptions because transfer time between operators does not add to the time efficiency of the system.

Because of different granularities and scales, there are several ways to describe a single application. Choosing the correct MoC and using this model correctly is vital to obtain efficient prototyping results. In following sections, SDF, CSDF, and hierarchical extensions of SDF are presented in more detail. The general theory of dataflow process networks can be found in [5]. This document concentrates on the models used in the study of LTE physical layer algorithms: SDF and its extensions.

3.2 Synchronous Data Flow

The Synchronous Dataflow (SDF) [10] is used to represent the behavior of an application at a coarse grain. An example of SDF graph is shown in Fig. 3.2a. SDF can model loops but not code behavioral modifications due to the nature of its input data. For example, an "if" statement cannot be represented in SDF.

The SDF model can be represented as a finite directed, weighted graph characterized by the graph $G = \langle V, E, d, p, c \rangle$ where:

- V is the `set of nodes`; each node represents an actor that performs computation on one or more input data streams and produces one or more output data streams.
- $E \subseteq V \times V$ is the `edge set`, representing channels which carry data streams.
- $d : E \rightarrow \mathbb{N}$ is the `delay` function with $d(e)$ the number of initial tokens on an edge e (represented by black dots on the graph).
- $p : E \rightarrow \mathbb{N}^*$ is the `production` function with $p(e)$ representing the number of data tokens produced by the e source actor at each firing and carried by e.
- $c : E \rightarrow \mathbb{N}^*$ is the `consumption` function with $c(e)$ representing the number of data tokens consumed from e by the e sink actor at each firing.

This graph is a `coordination language`, and so only specifies the topology of the network but does not give any information about the internal behavior of actors. The only behavioral information in the model is the amount of produced and consumed tokens. If only a simulation of the graph execution is needed, the actor behavior can usually be ignored and a few parameters like Deterministic Actor Execution Time (DAET) are necessary. However, if the model will be used to generate executable code, a host code must be associated with each actor.

3.2.1 SDF Schedulability

From a connected SDF representation, it should be possible to extract a valid single-core schedule as a finite sequence of actor firings. The schedulability property of the SDF is vital; this enables the creation of a valid multi-core schedule. A valid schedule can fire with no deadlock and its initial state is equal to its final state.

A SDF graph can be characterized by a matrix close to the incidence matrix in graph theory and called `topology matrix`. The topology matrix Γ is a matrix of size $|E| \times |V|$, in which each row corresponds to an edge e and each column corresponds to a node v. The coefficient $\Gamma(i, j)$ is positive and equal to N if N tokens are produced by the jth node on the ith edge. Conversely, the coefficient $\Gamma(i, j)$ is negative and equal to $-N$ if N tokens are consumed by the jth node on the ith edge.

Theorem 3.2.1 *A SDF graph is consistent if and only if* $\text{rank}(\Gamma) = |V| - 1$

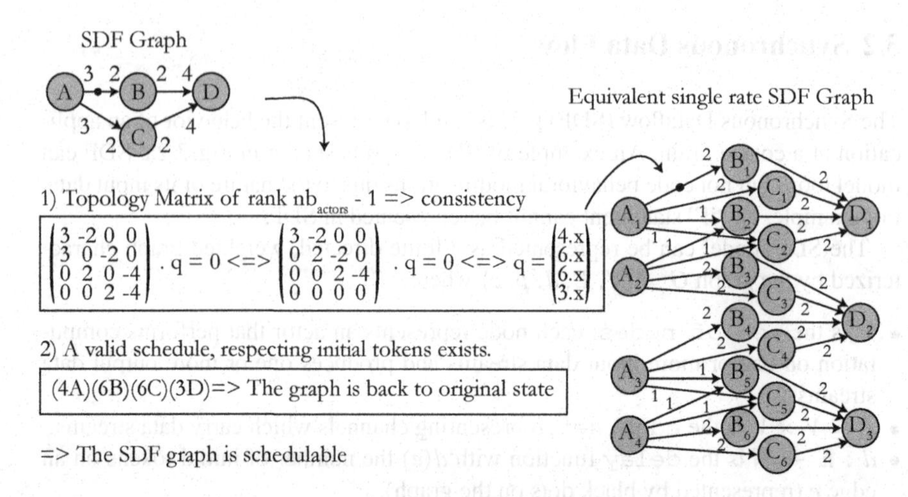

Fig. 3.3 The topology matrix and repetition vector of an SDF graph

Theorem 3.2.1 implies that there is an integer Basis Repetition Vector (BRV) q of size $|V|$ in which each coefficient is the repetition factor for the jth vertex of the graph in a schedule returning graph tokens in their original state. This basic repetition vector is the positive vector q with the minimal modulus in the kernel of the topology matrix such as $\Gamma.q = \{0\}$. Its computation is illustrated in Fig. 3.3. The BRV gives the repetition factor of each actor for a complete cycle of the network. Figure 3.3 shows an example of an SDF graph, its topology matrix and a possible schedule:

`Consistency` means that tokens cannot accumulate indefinitely in part of the graph [10, 11]. Consistency is a necessary (but not sufficient) condition for a SDF graph to be schedulable.

Theorem 3.2.2 *A SDF graph is schedulable if and only if it matches two conditions:*

1. *it is consistent,*
2. *a sequence of actor firing can be constructed, containing the numbers of firings from the basic repetition vector and respecting the actor firing rules (an actor can fire only if it has the necessary input tokens).*

The second condition can be called `deadlock freeness`. This condition is not respected when there are insufficient initial tokens and if an actor stays blocked (or deadlocked) indefinitely. The procedure to construct a single-core schedule to demonstrate schedulability is detailed in [12]. In Fig. 3.3, the chosen single-core scheduler respects Single Appearance Scheduling (SAS); a schedule where all instances of a given actor are grouped together. The single-core SAS schedule is optimal for code size. Other single-core schedules exist, where other parameters are optimized.

The ability to check graph schedulability before commencing algorithm and architecture matching is enormously advantageous. It means the algorithm graph is valid, regardless of the architecture on which it is mapped. The checking graph schedulability is one part of the total hardware and algorithm separation needed for a multi-core software development chain.

It may be noted that the above method is valid only for connected graphs. It cannot model several unconnected algorithms mapped on the same multi-core architecture. The case of a graph without enough edges to connect all vertices ($|E| < |V| - 1$) implies $\text{rank}(\Gamma) < |V| - 1$ and the graph is then considered non-schedulable. The "normal" case of a connected graph study implies $|E| \geq |V| - 1$. To obtain the Basis Repetition Vector and check schedulability, the equation $\Gamma.q = \{\mathbf{0}\}$ needs to be solved with q integer and $q \neq \mathbf{0}$. This equation can be solved using the Gauss algorithm as displayed in Algorithm 3.1. Using the Bachman-Landau asymptotic upper bound notation [13], the algorithm has a complexity of $O(|V|^2|E|)$ in non-trivial cases. It results in $O(|V|^2)$ divisions and $O(|V|^2|E|)$ matrix loads and store accesses.

Algorithm 3.1: Basis Repetition Vector Computation of Γ

Input: A topology matrix Γ of size $|E| \times |V|$
Output: A Basis Repetition Vector of size $|V|$ if $\text{rank}(\Gamma) = |V| - 1$, false otherwise

```
1  i = j = 1;
2  while i ≤ |E| and j ≤ |V| do
3      Find pivot (greatest absolute value) Γ_kj with k ≥ i;
4      if Γ_kj ≠ 0 then
5          Swap rows i and k;
6          Divide each entry in row i by Γ_ij;
7          for l = i + 1 to |E| do
8              Substract Γ_lj*row i from row l;
9          end
10         i=i+1;
11     end
12     j=j+1;
13 end
14 if Γ has exactly |V| − 1 non null rows then
15     Create VRB v = (1, 1, 1...);
16     for l = |V| − 1 to 1 do
17         Solve equation of raw l where only the rational v_l is unknown;
18     end
19     Multiply v by the least common multiple of v_i, 1 ≤ i ≤ |V|;
20     return v;
21 else
22     return false;
23 end
```

3.2.2 Single Rate SDF

The single rate Synchronous Dataflow model is a subset of the SDF where the productions and consumptions on each edge are equal. An example of single rate SDF graph is shown in Fig. 3.2b. It can thus be represented by the graph $G = \langle V, E, d, t \rangle$ where V, E and d are previously defined and $t : E \rightarrow N$ is a function with $t(e)$ representing the number of data tokens produced by the e source actor and consumed by the e sink actor at each firing. A SDF graph can be converted in its equivalent single rate SDF graph by duplicating each actor (Fig. 3.3). The number of instances is given by the BRV; the edges must be duplicated properly to connect all the single rate SDF actors.

An algorithm to convert a SDF graph to a single rate SDF graph is given in [14] p. 45. It consists of successively adding actors and edges in the single rate SDF graph and has a complexity $O(|V| + |E|)$ where V and E are the vertex set and the edge set respectively of the single rate SDF graph.

3.2.3 Conversion to a Directed Acyclic Graph

One common way to schedule SDF graphs onto multiple processors is to first convert the SDF graph into a precedence graph so that each vertex in the precedence graph corresponds to a single actor firing from the SDF graph. Thus each SDF graph actor A is "expanded into" q_A separate precedence graph vertices, where q_A is the component of the BRV that corresponds to A. In general, the precedence graph reveals more functional parallelism as well as data parallelism. A valid precedence graph contains no cycle and is called Directed Acyclic Graph (DAG) or Acyclic Precedence Expansion Graph (APEG). In a DAG, an actor without input edge is called `entry actor` and an actor without output edge is called `exit actor`. Unfortunately, the graph expansion due to the repeatedly counting each SDF node can lead to an exponential growth of nodes in the DAG. Thus, precedence-graph-based multi-core scheduling techniques, such as the ones developed in [15], generally have a complexity that is not bounded polynomially in the input SDF graph size, and can result in prohibitively long scheduling times for certain kinds of graphs [16] (Fig. 3.4).

Converting a single rate SDF graph into its corresponding DAG simply consists of ignoring edges with initial tokens. If the graph is schedulable, each of its cyclic paths will contain at least one initial token. Removing edges with initial tokens naturally breaks these cycles and creates a DAG. Using this precedence graph as the algorithmic input for scheduling results in inter-iteration dependencies not being taken into account. Converting a non-single rate SDF graph into its corresponding DAG is also possible but is a more complex operation as cycles must be unrolled before ignoring edges with initial tokens [17].

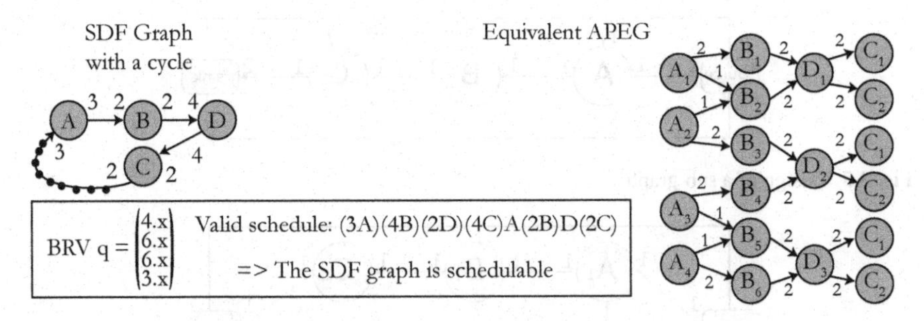

Fig. 3.4 SDF graph and its acyclic precedence expansion graph

3.3 Interface-Based Synchronous Data Flow

When designing an application, the designer may wish to use hierarchy to create independent graphs which may be instantiated independent of design. From a programmer view the hierarchy behaves as code closures since it defines limits for a portion of an application. This property of the hierarchy ensures that while a graph is instantiated, its behavior may not be modified by its parent graph, and that its behavior may not introduce deadlock in its parent graph. The defined composition rules ensure the graph to be deadlock free when verified, and can be also used to add hierarchy levels in a graph with no hierarchy in a bottom-up approach. Such a graph representation is depicted in Fig. 3.5.

To hierarchically design a graph in a top-down approach, the hierarchy semantic must ensure that the composed graph will have no deadlock when every level of hierarchy is independently deadlock free. To respect this rule, special nodes have been integrated in the proposed model thus restricting the hierarchy semantic. In the sections that follow a hierarchical vertex will refer to a vertex which embeds a sub-graph, and a sub-graph refers to the graph representing this hierarchy level.

3.3.1 Special Nodes

`Source node`: a Source node is a bounded source of tokens which represents the tokens available for one iteration of the sub-graph. This node behaves as an interface to the outside world. A source port is defined by the following four rules:

A-1 Source production homogeneity: a source node *Source* produces the same amount of tokens on all its outgoing connections $p(e) = n \quad \forall e \in \{Source(e) = Source\}$.

A-2 Interface Scope: the source node remains write-locked during an iteration of the sub-graph. This means that the interface can not be modified externally during the sub-graph execution.

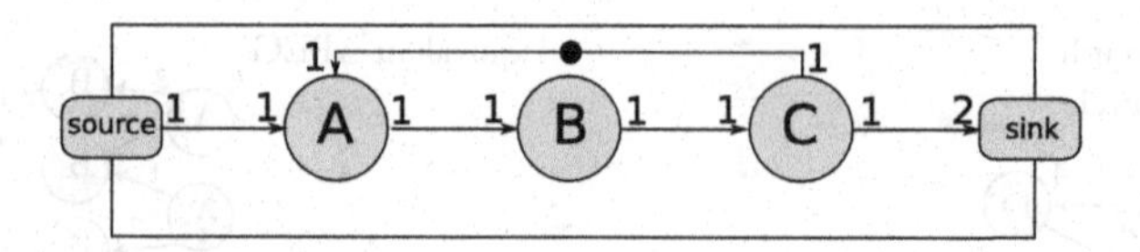

Fig. 3.5 Design of a sub-graph

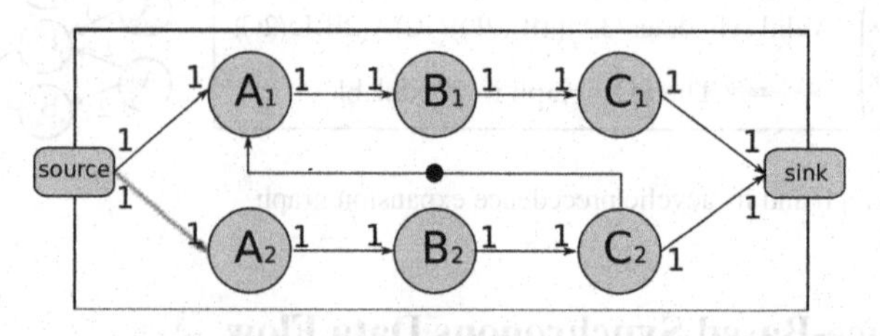

Fig. 3.6 A sub-graph after HSDF transformation

A-3 Interface boundedness: a source node cannot be repeated, thus any node consuming more tokens than made available by the node will consume the same token multiple times (ring buffer). $c(e)\%p(e) = 0 \;\; \forall e \in \{source(e) = source\}$.

A-4 SDF consistency: all the tokens made available by a source node must be consumed during an iteration of the sub-graph.

`Sink node`: a sink node is a bounded sink of tokens that represents the tokens to be produced by one iteration of the graph. This node behaves as an interface to the outside world. A sink node is defined by the four following rules:

B-1 Sink producer uniqueness: a sink node $Sink$ has only one incoming connection.

B-2 Interface scope: the sink node remains read-locked during an iteration of the sub-graph. This means that the interface cannot be read externally during the sub-graph execution.

B-3 Interface boundedness: A sink node cannot be repeated, thus any node producing more tokens than needed by the node will write the same token multiple times (ring buffer). $p(e)\%c(e) = 0 \;\; \forall e \in \{target(e) = Sink\}$.

B-4 SDF consistency: all tokens consumed by a sink node must be produced during one iteration of the sub-graph (Fig. 3.6).

3.3.2 Hierarchy Deadlock-Freeness

Consider a consistent connected SDF graph $G = \{g, z\}$, $g = \{Source, x, y, Sink\}$ where $Source$ is a source node and $Sink$ is a sink node, and z is an actor. In the following section the hierarchy rules described above are shown to respect the hierarchical vertex g so not to introduce deadlocks in graph G.

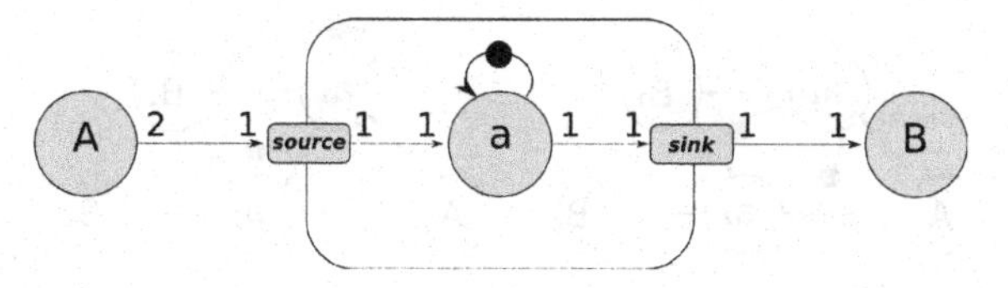

Fig. 3.7 Local edge in sub-graph

- If a simple path going from x to y containing more than one arc exists, this path cannot introduce cycle since there is at least one interface, meaning that the cycle gets broken. The user must account for this introduction of a cycle onto the actor g, by adding a delay to the top graph.
- Rules A2 - B2 ensure that all the data needed for one iteration of the sub-graph are available as soon as its execution starts, and that no external vertex can consume on the sink interface while the sub-graph is being executed. As a consequence external vertices strongly connected with hierarchical vertices may not be executed concurrently. The interface ensures the sub-graph content will be independent of external factors, as there is no edge $\alpha \in \left\{ \alpha' \| \left(\begin{array}{c} (src(\alpha') = x) \\ and \\ (snk(\alpha') \in C) \\ and \\ (snk(\alpha') \notin \{x, y\}) \end{array} \right) \right\}$ considering that $snk(\alpha') \notin \{x, y\})$ cannot happen.
- The design approach of the hierarchy means that there will never be a hidden delay. This is true even if there is a delay in the sub-graph, as an iteration of the sub-graph cannot start until its input interfaces are not fully populated with tokens.

These rules also guarantee that the edges of the sub-graph have a local scope, since the interfaces make the inner graph independent of external factors. So, when an edge that produced a cycle (contains a delay) in sub-graph that needs to be repeated, this iterating edge will not link multiple instances of the sub-graph. These rules are sufficient to ensure that a sub-graph will not create deadlocks when instantiated in a larger graph (Fig. 3.7).

3.3.2.1 Transformations

The main drawback of this representation is its behavior during hierarchy flattening. Previously defined rules restrict the edges of a sub-graph to local scope. This means that when removing the hierarchy care must be taken that no edges will propagate data among multiple instances of the sub-graph. By removing the hierarchy with the basic SDF semantic, local delays in the sub-graph may have significantly different interpretations when performing HSDF transformation on the graph (Fig. 3.8).

For this hierarchy semantic, prior to flattening a level of hierarchy, an HSDF transformation must be applied to preserve the edges scope of the sub-graphs.

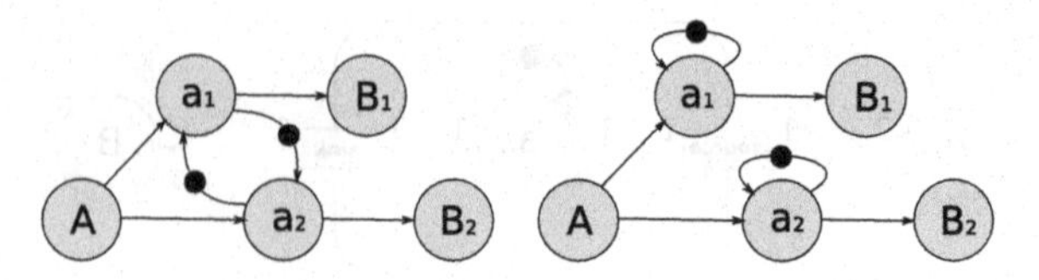

Fig. 3.8 Edge scope preservation in interface based hierarchy

It is also important to preserve interface behavior when removing the hierarchy. Interfaces can either behave as fork for Source port, when tokens need to be broadcasted to multiple nodes, or join when multiple tokens must be grouped. In order to avoid introducing deadlocks during the hierarchy flattening process, the interfaces must be replaced by a special vertex in the flatten graph. These special interfaces can then be treated to maximize the potential parallelism.

3.3.3 Hierarchy Scheduling

As pictured in [18] interfaces to the outside world must not be taken into account to compute the schedule-ability of the graph. For a hierarchy interpretation, internal interfaces have a meaning for the sub-graph, and so must be taken into account to compute the schedule-ability. In this way all the tokens on the interfaces can be guaranteed to be consumed/produced in an iteration of the sub-graph (see rules `A4-B4`).

Due to the nature of an interface, every connection coming/going from/to an interface must be considered as a connection to an independent interface. Adding an edge e to a graph G increases the rank of its topology matrix Γ (see Sect. 3.2.1) if the row added to Γ is linearly independent from the other row. Adding an interface to a graph G composed of N vertices, and one edge e connecting this interface to G adds a linearly independent row to the topology matrix. This increases the rank of the topology matrix of one, but adding the interface's vertex will yield in a $N+1$ graph: $rank(\Gamma(G_N)) = N-1 \Rightarrow rank(\Gamma(G_{N+1})) = rank(\Gamma(G_N))+1 = (N+1)-1$. The rank of the topology matrix remains equal to the number of vertices less one meaning that this graph remains schedule-able. In the case of the addition of an edge between a connected interface and any given vertex of the graph, the result is an edge between a newly created interface and the graph. Therefore, it can be seen from the proof above that schedule-ability will not be altered. This means that a sub-graph can be considered schedule-able if its actor graph (excluding interfaces) is schedule-able.

Before scheduling a hierarchical graph, every level of hierarchy must be verified to be deadlock free. Applying the balance equation to each level is sufficient to prove the absence of deadlocks for a level.

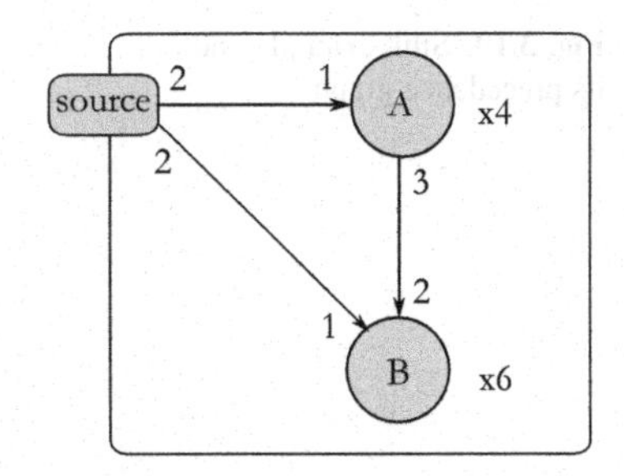

Fig. 3.9 Source example and its execution pattern

3.3.4 Hierarchy Behavior

Interface behavior can vary due to the graph schedule. This behavior variability can simplify the development process, but needs to be understood to ensure good representation of the applications.

3.3.4.1 Source Behavior

As defined in the *Source* interface rules, a *Source* interface can have multiple outgoing (independent) connection. Care must be taken as reading more tokens than made available results in reading modulo the number of tokens available (circular buffer). This means that the interface can behave like a broadcast. In Fig. 3.9, vertices A and B have to execute 4 and 6 times respectively to consume all data provided by the port. In this example, the *Source* interface will broadcast twice to vertex A and three times to vertex B. Hence the designer must always be aware that the interfaces can affect the inner graph schedule.

3.3.4.2 Sink Behavior

As defined in the *Sink* interface rules, a *Sink* interface can have only one incoming connection, and writing more tokens than needed on the interface results in writing modulo the number of tokens needed (circular buffer). In Fig. 3.10, the vertex B writes 3 tokens in a *Sink* that consumes only one token, due to the circular buffer behavior, so only the last token written will be made available to the outside world. This behavior permits the design of iterative patterns without increasing the number of edges. This behavior can also produce design errors; when no precedence exists between multiple occurrences of a vertex that writes to an output port, a parallel execution of these occurrences leads to a concurrent access to the interface. As a consequence, this will produce indeterminate data in the *Sink* node. This also leads to dead codes from the node occurrences that do not write to the *Sink* node.

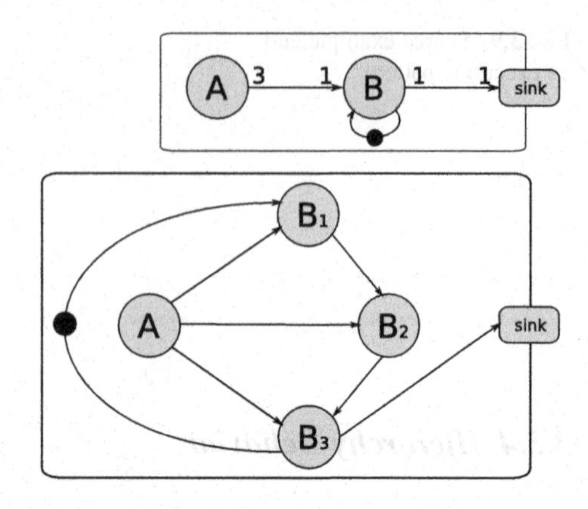

Fig. 3.10 Sink example and its precedence graph

3.3.5 Hierarchy Improvements

This introduction of hierarchy eases the work of the designer, as it allows independent subsystem design which permits instantiation in any application. Furthermore, hierarchy improves the application representation when using the same constraints as the clustering techniques (scheduling complexity reduction, buffer requirements...). These improvements are based on the initial choices of the designer but may be refined by automatic transformation, which allows more performance to be extracted from the graph.

3.4 Cyclo Static Data Flow

The Cyclo Static Dataflow (CSDF) model is introduced in [12] and an example graph is shown in Fig. 3.2e. The CSDF model cannot express more algorithms than SDF but it can express certain algorithms in a reduced way. CSDF can also enhance parallelism and reduce memory necessary for execution. CSDF is used in the study of LTE uplink and downlink streams. The token production and consumption of each actor can vary over time, following a periodic form statically chosen. For example, an actor can consume one token, then two tokens, then one token again, and so on. In this example, the actor has two phases. In CSDF, token productions and consumptions are patterns instead of scalars. A SDF model can be represented as a finite directed, weighted graph characterized by the graph $G = \langle V, E, d, \phi, p, c \rangle$ where:

- V is the `set of nodes`; each node representing an actor.
- $E \subseteq V \times V$ is the `edge set`, representing data channels.
- $d : E \rightarrow \mathbb{N}$ is the `delay` function with $d(e)$ the number of initial tokens on an edge e.

- $\phi : V \to \mathbb{N}$) is the phase function with $\phi(v)$ the number of phases in the pattern execution of v.
- $p : E \times \mathbb{N} \to \mathbb{N}^*$ is the production function with $p(e, i)$ representing the number of data tokens produced by the e source actor at each firing of phase i and carried by e.
- $c : E \times \mathbb{N} \to \mathbb{N}^*$ is the consumption function with $c(e, i)$ representing the number of data tokens consumed from e by the e sink actor at each firing of phase i.

The CSDF model introduces the notion of actor state; an actor does not behave the same at each firing. However, the fixed pattern of execution permits to check the graph schedulability at compile-time.

3.4.1 CSDF Schedulability

Like a SDF graph, a CSDF graph can be characterized by its topology matrix. The topology matrix Γ is the matrix of size $|E| \times |V|$, where each row corresponds to an edge e and each column corresponds to a node v. The coefficient $\Gamma(i, j)$ is positive and equal to $\sum_{k=1}^{\phi(j)} p(i, k)$ if a pattern $p(i, k)$ is produced by the jth node on the ith edge. Conversely, the coefficient $\Gamma(i, j)$ is negative and equal to $-\sum_{k=1}^{\phi(j)} c(i, k)$ if a pattern $c(i, k)$ is consumed by the jth node on the ith edge.

The topology matrix gathers the cumulative productions and consumptions of each actor during its complete period of execution. A CSDF graph G is consistent if and only if the rank of its topology matrix Γ is one less than the number of nodes in G (Theorem 3.2.1). However, the smallest vector q' in Γ null space is not the BRV. q' only reflects the number of repetitions of each complete actor cycle and must be multiplied for each actor v by its number of phases $\phi(v)$ to obtain the BRV q:

$$q = \Phi \times \mathbf{q}', \quad \text{with} \quad \Phi = \text{diag}(\phi(v_1), \phi(v_2), \ldots, \phi(v_m)) \tag{3.1}$$

A valid sequence of firings with repetitions given by the BRV is still needed to conclude the schedulability of the graph. The process is illustrated in Fig. 3.11. It can be concluded on CSDF schedulability as in the SDF case. Fig. 3.12 illustrates a compact representation of a CSDF graph and its SDF equivalent. In this simple example, the number of vertices has been reduced by a factor 2.3 by using CSDF instead of SDF.

3.5 Dataflow Hierarchical Extensions

A hierarchical dataflow model contains hierarchical actors. Each hierarchical actor contains a net of actors. There are several ways to define SDF actor hierarchy. The first SDF hierarchy was that of clustering described in [16] by Pino, et al. The study

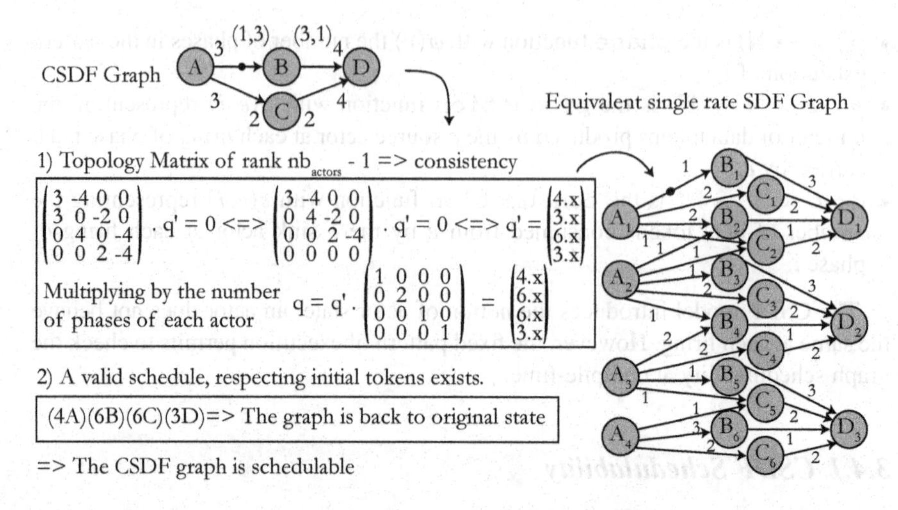

Fig. 3.11 The topology matrix and repetition vector of a CSDF graph

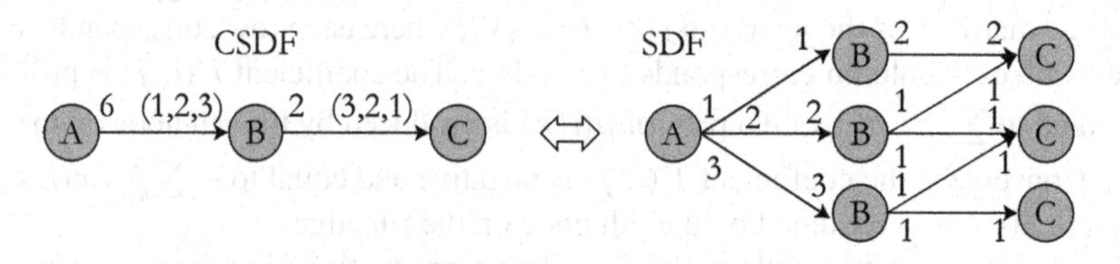

Fig. 3.12 Illustrating compactness of CSDF model compared to SDF

takes a bottom-up approach and defines rules that extract the actors from a complex graph which can be clustered within a coarser-grain actor. The primary advantage of this technique is that clustered actors are then easier than the individual actors in scheduling process manipulations.

However, the type of hierarchy that is of interest in this document is a top-down one: a SDF hierarchy that allows a designer to compose sub-graphs within an upper-level graph. Two such hierarchy models have been defined and both will be studied in this document: the Parameterized SDF (PSDF [19]) and the Interface-Based Hierarchical SDF (IBSDF [20]).

3.5.1 Parameterized Dataflow Modeling

Parameterized dataflow modeling is a technique that can be applied to all dataflow models with deterministic token rates, in particular to SDF and CSDF [21]. It consists of creating a hierarchical view of the algorithm whose subsystems are then composed of 3 graphs:

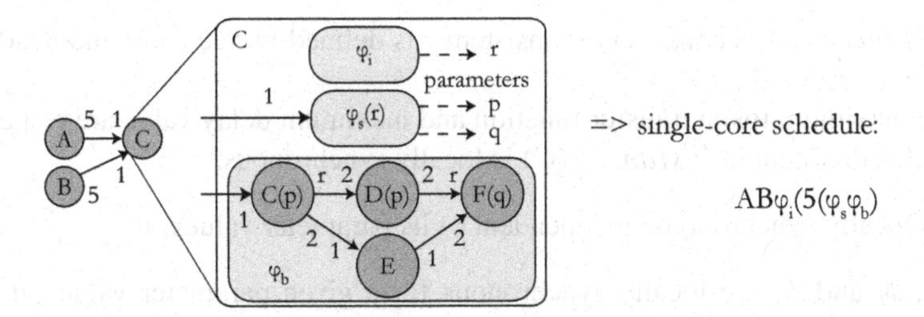

Fig. 3.13 PSDF example

- the body graph ϕ_b: this is the effective dataflow sub-graph. Its productions and consumptions depend on certain parameters. These parameters are called topology parameters because they affect the topology of the body graph. The behavior of the actors (not expressed in the coordination language) depends on parameters that will be called actor parameters.
- the init graph ϕ_i: this graph is launched one time only when a subsystem interface is reached, i.e. once per `invocation` of the subsystem. It performs a reconfiguration of the subsystem, setting parameters that will be common to all ϕ_b firings during the invocation. ϕ_i does not consume any data token.
- the subInit graph ϕ_s: it is launched one time before each body graph firing and also sets parameters. ϕ_s can consume data tokens.

Parameterized dataflow modeling is generally applied to SDF graphs. In this case, it is called PSDF for Parameterized Synchronous Dataflow. A complete description of PSDF can be found in [19]; only the properties of interest to LTE are studied here. Figure 3.13 shows a small PSDF top graph G with three actors A, B, and C. A and B are atomic SDF actors, i.e. they contain no hierarchical subsystem. C contains a PSDF sub-graph with one topology parameter r and two actor parameters p and q. A compressed view of a single-core schedule for G is also displayed in the Figure. The internal schedule of ϕ_b is not shown because this schedule depends on the topology parameter r alone.

PSDF can also be seen as a dynamic extension of SDF because token productions and consumptions depend on topology parameters that can vary at run-time. The PSDF parameter values must be restricted to a domain (a set of possible values) to ensure that the graph can be executed with a bounded memory. The set of the graph G parameter domains is written $DOMAIN(G)$. Similarly, the token rate at each actor port is limited to a maximum token transfer function and the delays on each edge are limited to a maximum delay value.

The most interesting property of PSDF is that its hierarchical form enables the compile-time study of local synchrony in the subsystem [19]. Local synchrony ensures that each local synchronous subsystem of a PSDF graph can be scheduled in SDF fashion: either switching between pre-computed SDF schedules or scheduling at run-time. For a given set of parameters values C, the graph $instance_G(C)$ is locally synchronous if:

1. $instance_G(C)$ is consistent (consistency is defined in Sect. 3.2) and deadlock-free,
2. the maximum token transfer function and maximum delay value are respected,
3. each subsystem in $instance_G(C)$ is locally synchronous.

G is locally synchronous (independent of its parameter values) if:

1. ϕ_b, ϕ_i and ϕ_s are locally synchronous for a given parameter value set $C \subset DOMAIN(G)$,
2. ϕ_i and ϕ_s produce one parameter token per firing,
3. the number of dataflow input tokens of ϕ_s is independent of the general dataflow input of the subsystem,
4. the productions and consumptions at interfaces of ϕ_b do not depend on parameters configured in ϕ_s.

If G is locally synchronous, a local SDF schedule can be computed at run-time to execute the graph. Though ϕ_i and ϕ_s can set both topology and actor parameters, general practice is for ϕ_s to only change actor parameters so identical firings of ϕ_b occur for each subsystem invocation. When this condition is satisfied, a local SDF schedule only needs to be computed once after ϕ_i is executed. Otherwise, the local SDF schedule needs to be computed after each ϕ_s firing.

In this LTE study, Parameterized CSDF is also of interest; this is equivalent to PSDF except the parameters are not only scalars but can be patterns of scalars similar to the CSDF model. For this document, the PDSF init and the PSDF sub-init graphs employed are degenerated graphs containing only a single actor and do not consume or produce any data tokens. They are thus represented by a simple rectangle illustrating the actor execution (Chap. 8).

3.5.2 *Interface-Based Hierarchical Dataflow*

Interface-Based Hierarchical Dataflow Modeling is introduced by Piat et al. in [20, 22]. Like Parameterized Dataflow Modeling, it may be seen as a meta-model that can be applied to several deterministic models. The objective of Interface-Based Hierarchical Dataflow Modeling is to allow a high-level dataflow graph description of an application, composed of sub-graphs. If an interface is not created to insulate the upper graph from its sub-graphs, the actor repetitions of sub-graphs can result in the actor repetitions of the upper graph appearing illogical to the programmer. The Interface-Based Hierarchical Dataflow Modeling has been described in combination with SDF in the so-called Interface-Based Hierarchical SDF or IBSDF (see Sect. 3.3). The model obtained encapsulates sub-graphs and protects upper graphs. Encapsulation is a very important model property; it enables the combination of semiconductor intellectual property cores (commonly called IPs) in electronic design and is the foundation of object oriented software programming.

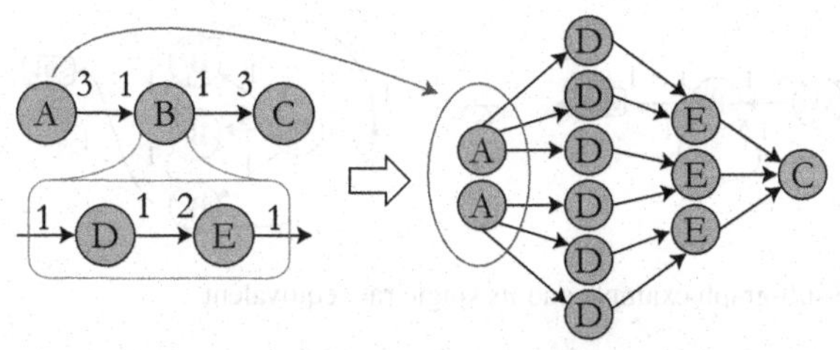

(a) Hierarchical SDF Example Flattening

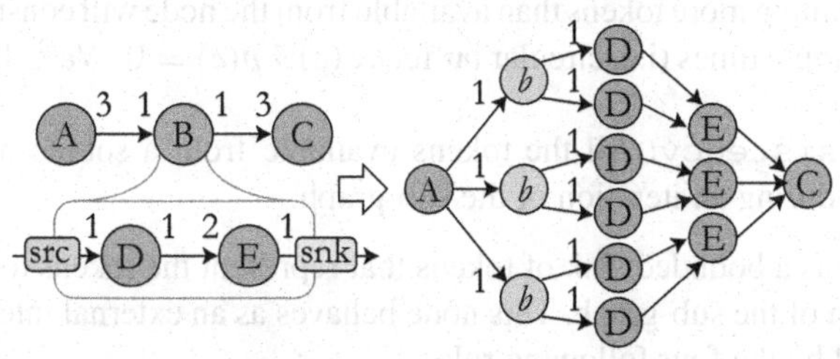

(b) IBSDF Example Graph Flattening

Fig. 3.14 Comparing direct hierarchical SDF with IBSDF

The operation of hierarchy flattening consists of including a sub-graph within its parent graph while maintaining the general system behavior. Figure 3.14a shows an example of a SDF graph with a hierarchical actor of which the programmer will probably want the sub-graph to be repeated three times. Yet, the sub-graph behavior results in a very different flattened graph, repeating the actor *A* from the upper graph. Figure 3.14b illustrates the same example described in as a IBSDF graph. Source and sink interfaces have been introduced. The resulting flattened graph shape is more intuitive. The additional broadcast actors have a circular buffering behavior: as each input is smaller in size than the sum of their outputs, they duplicate each input token in their output edges.

Compared with other SDF actors, the IBSDF interfaces have specific behavior. They are designed to constitute code closures (i.e. semantic boundaries). The IBSDF type of hierarchy is closer to host language semantics such as C, Java and Verilog. The additional features have been shown to introduce no deadlock [20].

A Source node is a bounded source of tokens which represents the tokens available for a sub-graph iteration. This node acts as an interface to the outside world. A source port is defined by the four following rules:

1. `Source production homogeneity`: A source node *Source* produces the same amount of tokens on all its outgoing connections $p(e) = n \;\; \forall e \in \{Source(e) = Source\}$.
2. `Interface Scope`: The source node remains write-locked during an iteration of the sub-graph. This means that the interface cannot be externally filled during the sub-graph execution.

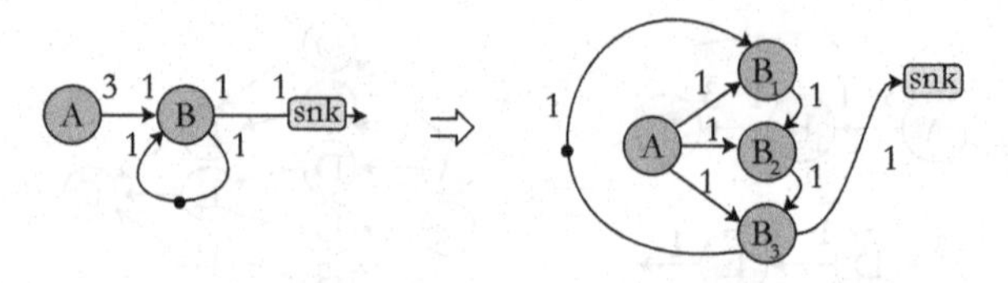

Fig. 3.15 IBSDF sub-graph example and its single rate equivalent

3. `Interface boundedness`: A source node cannot be repeated, thus any node consuming more tokens than available from the node will consume the same tokens multiple times (i.e. circular buffer). $c(e)\%p(e) = 0 \quad \forall e \in \{source(e) = source\}$.
4. `SDF consistency`: All the tokens available from a source node must be consumed during an iteration of the sub-graph.

A sink node is a bounded sink of tokens that represent the tokens to be produced by one iteration of the sub-graph. This node behaves as an external interface. A sink node is defined by the four following rules:

1. `Sink producer uniqueness`: A sink node $Sink$ only has one incoming connection.
2. `Interface Scope`: The sink node remains read-locked during an iteration of the sub-graph. This means that the interface cannot be externally read during the sub-graph execution.
3. `Interface boundedness`: A sink node cannot be repeated, thus any node producing more tokens than needed will write the same tokens multiple times (i.e. ring buffer). $p(e)\%c(e) = 0 \quad \forall e \in \{target(e) = Sink\}$.
4. `SDF consistency`: All tokens consumed by a sink node must be produced during an iteration of the sub-graph.

Source and sink nodes cannot be repeated, providing a fixed reference interface to combine these nodes in an upper level graph. A source node, behaving as a circular buffer, sends the same tokens to its graph possibly several times. A sink node, behaving as a circular buffer, stores the last tokens received from its graph. Figure 3.15 shows an example of an IBSDF graph and its single rate equivalent. IBSDF has an extremely important property: an IBSDF graph is schedulable if and only if its constituent SDF sub-graphs are schedulable [22]. Moreover, the interfaces do not influence SDF sub-graph schedulability. Independently applying the SDF BRV calculation to each algorithm part with complexity $O(|V|^2|E|)$ presented in Sect. 3.2 makes the overall calculation much faster.

In this document, IBSDF is the model chosen to describe LTE algorithms for rapid prototyping. Several features have been added for the use of IBSDF in PREESM to enhance expressivity and ease code generation. The final model is presented in Sect. 6.2.1.

While Chaps. 2 and 3 introduced the target application of this study and the MoCs available to model it, Chap. 4 will detail the process of rapid prototyping, which will be applied to LTE in the following chapters.

References

1. Chang WT, Ha S, Lee EA (1997) Heterogeneous simulation—mixing discrete-event models with dataflow. J VLSI Signal Process 15(1):127–144
2. International Technology Roadmap for Semiconductors (2009) I.T.R.: Design. http:www.itrs.net
3. Sgroi M, Lavagno L, Sangiovanni-Vincentelli A (2000) Formal models for embedded system design. IEEE Des Test Comput 17(2):14–27
4. Sgroi M, Lavagno L, Watanabe Y (1999) Sangiovanni-Vincentelli, A.: Synthesis of embedded software using free-choice petri nets. In: Proceedings of the 36th annual ACM/IEEE design automation conference, pp 805–810
5. Lee EA, Parks TM (1995) Dataflow process networks. Proc IEEE 83(5):773–801
6. Kahn G (1974) The semantics of a simple language for parallel programming. Inf process 74:471–475
7. Janneck JW, Mattavelli M, Raulet M, Wipliez M (2010) Reconfigurable video coding: a stream programming approach to the specification of new video coding standards. In: Proceedings of the first annual ACM SIGMM conference on, Multimedia systems, pp 223–234
8. Nezan JF (2002) Integration de services video mpeg sur architectures paralleles. Ph.D. thesis, IETR INSA Rennes
9. Sinnen O (2007) Task scheduling for parallel systems (Wiley series on parallel and distributed computing). Wiley-Interscience, Hoboke
10. Lee E, Messerschmitt D (1987) Synchronous data flow. Proc IEEE 75(9):1235–1245
11. Buck JT, Lee EA (1993) Scheduling dynamic dataflow graphs with bounded memory using the token flow model. In: 1993 IEEE international conference on acoustics, speech, and, signal processing, 1993 ICASSP-93, vol. 1
12. Bilsen G, Engels M, Lauwereins R, Peperstraete JA (1995) Cyclo-static data flow. In: International conference on Acoustics, speech, and signal processing, 1995, vol. 5
13. Cormen TH, Leiserson CE, Rivest RL, Stein C (2001) Introduction to algorithms, 2nd edn. MIT Press, Cambridge
14. Sriram S, Bhattacharyya SS (2009) Embedded multiprocessors: scheduling and synchronization, 2nd edn. CRC Press, Boca Raton
15. Kwok Y (1997) High-performance algorithms of compile-time scheduling of parallel processors. Ph.D. thesis, Hong Kong University of Science and Technology
16. Pino JL, Bhattacharyya SS, Lee EA (1995) A hierarchical multiprocessor scheduling framework for synchronous dataflow graphs. Laboratory, University of California at Berkeley pp 95–36. doi:10.1.1.24.3759. http://citeseerx.ist.psu.edu/viewdoc/summary?doi=?doi=10.1.1.24.3759
17. Piat J (2010) Data flow modeling and multi-core optimization of loop patterns. Ph.D. thesis, INSA Rennes
18. Lee EA, Messerschmitt DG (1987) Static scheduling of synchronous data flow programs for digital signal processing. IEEE Trans Comput 36(1):24–35
19. Bhattacharya B, Bhattacharyya S (2001) Parameterized dataflow modeling for DSP systems. IEEE Trans Signal Process 49(10):2408–2421
20. Piat J, Bhattacharyya SS, Pelcat M, Raulet M (2009) Multi-core code generation from interface based hierarchy. DASIP 2009
21. Bhattacharya B, Bhattacharyya S (2002) Consistency analysis of reconfigurable dataflow specifications. In. Lecture notes in computer science, pp 308–311
22. Piat J, Bhattacharyya SS, Raulet M (2009) Interface-based hierarchy for synchronous data-flow graphs. SAMOS conference IX

Chapter 4
Rapid Prototyping and Programming Multi-Core Architectures

4.1 Introduction

This chapter gives an over view of the existing work on rapid prototyping and multi-core deployment in the signal processing world. The concept of rapid prototyping was introduced in Fig. 1.2 when outlining the structure of this document. It consists of automatically generating a system simulation or a system prototype from quickly constructed models. Rapid prototyping may be used for several purposes; this study uses it to manage the parallelism of DSP architectures. Parallelism must be handled differently for the macroscopic or microscopic views of a system. The notions that are developed in this section will be used to study LTE in Part II. Section 4.2 gives an insight into embedded heterogeneous architectures. Section 4.3 is an overview of the multi-core programming techniques and Sect. 4.4 focuses on the internal mechanism of multi-core scheduling. Finally, Sect. 4.5 presents some results from the literature on automatic code generation.

4.1.1 The Middle-Grain Parallelism Level

The rapid prototyping and programming method focuses on particular classes of algorithms and architecture, and a certain degree of parallelism. This study targets signal processing applications running on multi-core heterogeneous systems and the objective is to exploit their middle-grain parallelism. These terms are defined in next sections.

The problem of parallel execution is complex and analysis is aided if it is broken down into sub-problems. An application does not behave similarly at system-level or low-level. Three levels of granularity are usually evoked in literature and concurrency can be exploited at each of these three levels: instruction-Level Parallelism (ILP), Thread-Level Parallelism (ThLP) and Task-Level Parallelism (TLP). An `instruction` is a single operation of a processor defined by its Instruction

M. Pelcat et al., *Physical Layer Multi-Core Prototyping*,
Lecture Notes in Electrical Engineering 171, DOI: 10.1007/978-1-4471-4210-2_4,

Table 4.1 The Levels of parallelism and their present use

Parallelism level	Target architectures	Source code	Parallelism extraction
Instruction-level	SIMD or VLIW architectures	Imperative or functional programming languages (C, C++, Java...)	Parallel optimizations of a sequential code by the language compiler
Thread level	Multi-core architecture with multi-core RTOS	Multi-threaded programs, parallel language extensions...	Compile-time or run-time partitioning and scheduling under data and control dependency constraints
Task level	Multi-core architecture with or without multi-core RTOS	Loosely-coupled processes from any language	Compile-time or run-time partitioning and scheduling

Set Architecture (ISA). The term `task` is used here in the general sense of program parts that share no or little information with each other. Tasks can contain several `threads` that can run concurrently but have data or control dependencies. The three levels of granularity are used to specify parallelism in a hierarchical way. Table 4.1 illustrates the current practices in parallelism extraction of applications.

The `Task Level Parallelism` concerns loosely coupled processes with few dependencies. Processes are usually fairly easy to execute in parallel. Processes can either be assigned to a core manually or by a Real-Time Operating System (RTOS). The few shared resources can usually be efficiently protected manually.

The `Instruction-Level Parallelism` has been subject of many studies in the 1990s and 2000s and effective results have been obtained, either executing the same operation simultaneously (Single Instruction Multiple Data or SIMD) or even different independent operations simultaneously (Very Long Instruction Word or VLIW [1]) on several data sets. Instruction-Level Parallelism can now be extracted automatically by compilers from sequential code, for instance written in C, C++ or Java. The most famous SIMD instruction set extensions are certainly the MMX and Streaming SIMD Extensions (SSE) x86. The Texas Instruments c64x and c64x+ Digital Signal Processor (DSP) cores have VLIW capabilities with 8 parallel execution units (Sect. 5.1.1). There is a natural limit to such parallelism due to dependencies between successive instructions. Core instruction pipelines complete the SIMD and VLIW parallelism, executing simultaneously different steps of several instructions. Instruction-level parallelizing and pipelining are now at this limit and it is now the middle-grain parallel parallelism functions with some data dependency that need further study: the Thread-Level Parallelism (ThLP).

`Thread Level Parallelism` is still imperfectly performed and multi-threading programming may not be a sustainable solution, due to its non-predictability in many cases, its limited scalability and its difficult debugging [2]. Yet, ThLP is the most important level for parallelizing dataflow applications on the current multi-core architectures including a small number of operators because ThLP usually contains much parallelism with reduced control, i.e. "stable" parallelism. The term operator will be used to designate both software cores and hardware Intellectual Properties (IP), the latter being dedicated to one type of processing (image processing, Fourier transforms, channel coding...). In past decades, the studies on operating systems have focused on the opposite of thread parallelization: the execution of concurrent threads on sequential machines, respecting certain time constraints. The concept of thread was invented for such studies and may not be suited for parallel machines. The solution to the parallel thread execution problem possibly lies in code generation from dataflow models. Using dataflow models, threads are replaced by actors that have no side effects, are purely functional and communicate through well defined data channels [3]. A side effect is a modification of a thread state by another thread due to shared memory modifications. Using actors, thread-Level parallelism then becomes `Actor Level Parallelism` (ALP); the general problem of managing parallelism at ThLP and ALP will be called `middle-grain parallelism`.

While task and instruction parallelism levels are already managed efficiently in products, middle-grain parallelism level is still mostly at a research phase. The PREESM framework aims to provide a method for an efficient and automatic middle-grain parallelization.

4.2 Modeling Multi-Core Heterogeneous Architectures

4.2.1 Understanding Multi-Core Heterogeneous Real-Time Embedded DSP MPSoC

The target hardware architectures of this work are associated with a great number of names and adjectives. These terms need definitions:

- `DSP` can refer to both Digital Signal Processing and Digital Signal Processor. In this document, DSP is used to refer to Digital Signal Processors, i.e. processors optimized to efficiently compute digital signal processing tasks.
- An `embedded system` is a calculator contained in a larger system embedded into the environment it controls. Real-time embedded systems include mobile phones, multimedia set-top boxes, skyrocket and, of interest in this study, wireless communication base stations. Only systems with very strict constraints on power consumption will be considered; this does not include hardware architectures consuming tens or hundreds of Watts such as Graphical Processing Units (GPU) or general purpose processors. The study concentrates on Digital Signal Processors (DSP), which are well suited for real-time embedded signal processing systems.
- A `real-time system` must respect the time constraints defined by the running application and produce output data at the exact time the environment requires it. Real-time systems include video or audio decoding for display, machine-tool control, and so on.

- `Multi-Core` is a generic term for any system with several processor cores able to concurrently execute programs with Middle-grain or Task-Level Parallelism.
- `Heterogeneous` refers to the existence of different types of operators within the system. It can also refer to the non-symmetrical access to resources, media with differing performances or cores with different clock speeds.
- A `Multi-Processor System-on-Chip` or MPSoC is a term widely used to designate a multi-core system embedded in a single chip.

In general, modern multi-core embedded applications are manually programmed, using C or C++ sequential code, and optimized using assembly code. Manual multi-core programming is a complex, error-prone and slow activity. The International Technology Roadmap for Semiconductors (ITRS) evaluates that embedded software productivity doubles every five years; this is a much slower growth than that of processor capabilities. ITRS predicts the cost of embedded software will increase until 2012 where it will reach three times the total cost of embedded hardware development. At that time, parallel development tools are predicted to reduce software development costs drastically [4]. The goal of this study is to contribute to this evolution.

Today, Multi-core digital signal operators, including up to several DSPs, face a choice. They may choose to use a small number of complex cores or to use many simple ones. The more cores are involved, the more complex the problem of exploiting them correctly becomes. With a few exceptions, the present trend in powerful digital signal processing architectures is to embed a small number of complex processor cores on one die [5]. Software-defined radio, where signal processing algorithms are expressed in software rather than in hardware, is now widely used for embedded systems including base stations. However, certain repetitive compute-hungry functionalities can make a totally software-defined system suboptimal. Hardware accelerators, also known as coprocessors, can process compute-hungry functionalities in a power-efficient way. They cost programming flexibility but increase the system power efficiency. The current work studies heterogeneous multi-core and multi-processor architectures with a few (or a few tens) complex DSP cores associated with a small number of hardware coprocessors. Architectures with more than a few tens of operators are (currently) called "many-core".

In following sections, the term multi-core system is used for the very general case of either one processor with several cores or several interconnected processors. The next section explores architecture modeling for such systems.

4.2.2 Literature on Architecture Modeling

Today, processor cores are still based on the `Von Neumann model` created in 1946, as illustrated in Fig. 4.1. It consists of a memory containing data and programs, an Arithmetic and Logic Unit (ALU) that processes data, input and output units that communicate externally, and a control unit that drives the ALU and man-

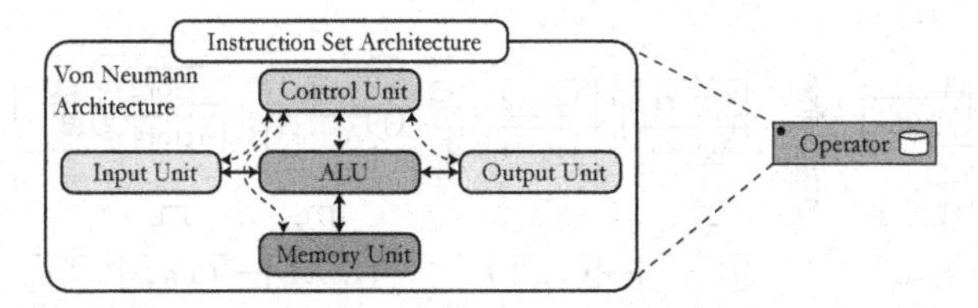

Fig. 4.1 The Von Neumann architecture of an operator

ages the program. Externally, this micro-architecture can be seen as a black box implementing an `Instruction Set Architecture` (ISA), writing/reading external memory and sending/receiving messages. Such an entity is called an operator. An operator can have an ISA so restricted that it is no longer considered to a processor core but a coprocessor dedicated to one or a few tasks.

The literature on architecture models is less extensive than that of algorithm models. The majority of the literature focuses on classifying architectures, and associating specific architectures to categories. Processors have been split between Complex Instruction Set Computer (CISC) and Reduced Instruction Set Computer (RISC) depending on their instruction set complexity [6]. The limit between CISC and RISC machines is generally unclear. Many instruction sets exist, each with its own tradeoff between core clock speed and advanced instruction capabilities. Tensilica [7] is a company known for tools capable of generating DSP cores with customizable ISA. Thus, CISC or RISC may no longer be a valid criterion to define a processor.

The Flynn taxonomy [8] has introduced four categories of machines classified depending on their instruction and/or data parallelism at a given time: Single Instruction Single Data (SISD), Single Instruction Multiple Data (SIMD), Multiple Instruction Single Data (MISD) and Multiple Instruction Multiple Data (MIMD). MIMD machines, requiring many control information at each clock cycle, have been implemented using Very Long Instruction Words (VLIW). MIMD systems have a sub-classification [9] based on their memory architecture types, splitting MIMD machines into 3 categories:

- A `Uniform Memory Access` (UMA) machine has a memory common to all its operators which is accessed at the same speed with the same reliability. This shared memory is likely to become the bottleneck of the architecture; two solutions can increase the access speed: it may be divided into banks attributed to a specific operator in a given time slot or it may be accessed via a hierarchical cache.
- A `Non Uniform Memory Access` (NUMA) machine has a memory common to all its operators but access speed to this memory is heterogeneous. Some operators have slower access speeds than others.
- A `NO Remote Memory Access` (NORMA) machine does not have a shared memory. Operators must communicate via messages. NORMA architecture is usually called `distributed architecture` because no centralized memory entity exists.

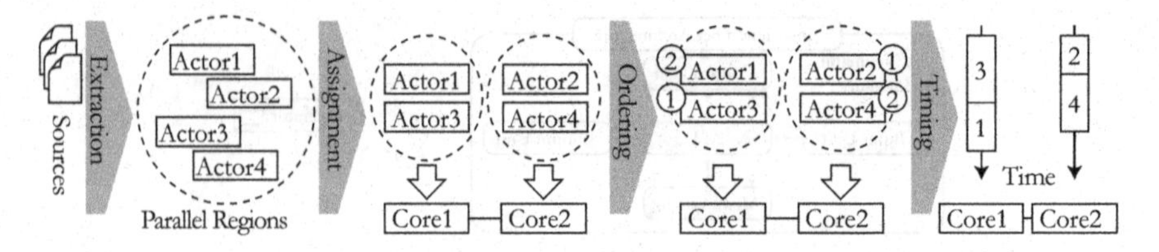

Fig. 4.2 The four phases of middle-grain parallelization: extraction, assignment, ordering and timing

In Chap. 5, a new system-level architecture model is introduced. Architecture models exist in the literature but they were not suited to model target architectures for rapid prototyping. The SystemC language [10] and its Transaction Level Modeling (TLM) is a set of C++ templates that offers an efficient standardized way to model hardware behavior. Its focus is primarily the study and debugging of the hardware and software functionalities before the hardware is available. It defines abstract modules that a programmer can customize to its needs. For this rapid prototyping study, more precise semantics of architecture element behaviors are needed than those defined in SystemC TLM. Another example of language targeting hardware simulation is a language created by the Society of Automotive Engineers (SAE) and named Architecture Analysis and Design Language (AADL [11]). The architecture part of this language is limited to the three nodes: processor, memory and bus. Processes and threads can be mapped onto the processors. This model is not adaptable to the study presented here because it focuses on imperative software containing preemptive threads and the architecture nodes are insufficient to model the target architectures with Direct Memory Accesses (DMA) and switches (Sect. 5.1.1).

In next section, existing work on multi-core programming is presented.

4.3 Multi-Core Programming

4.3.1 Middle-Grain Parallelization Techniques

Middle-grain parallelization consists of four phases: `extraction`, `assignment`, `ordering` and `timing`. Parallel actors must be extracted from the source code or model and all their dependencies must be known. Parallel actors are then assigned (or mapped) to operators available in the target architecture. Their execution on operators must also be ordered (or scheduled) because usually, there are fewer operators than actors. Finally, a start time for the execution of each actor must be chosen. The four tasks of extraction, assignment, ordering and timing, are illustrated in Fig. 4.2. Graham showed in the 1960s that relaxing certain constraints in a parallel system with dependencies can lead to worsened performance when using a local mapping and scheduling approach [12], making middle-grain parallelism extraction a particularly complex task.

In [13], Park et al. divide the current embedded system design methods into 4 categories based on their input algorithm description. All four methods focus on the middle-grain parallelism level:

- `Compiler-based` design methods extract middle-grain parallel actors from sequential code such as C-code, and map actors on different operators to enable their parallel execution. The method is analogous to that broadly adopted for the Instruction-Level. The obvious advantage of this method is the automatic parallelization of legacy code and, for an efficient compiler, no further effort is required from software engineers to translate their code. The key issue in the complex operation of parallelism extraction is to find middle-grain actors, coarse-grained enough to group data transfers efficiently and fine-grained enough to enable high parallelism. Using imperative and sequential code as an input, these methods are very unlikely to exploit all the algorithm potential parallelism. An example of a compiler implementing compiler-based method is the MPSoC Application Programming Studio (MAPS) from Aachen RWTH University [14] which generates multi-threaded C code from sequential C code and parallelizes these threads.
- `Language-extension` design methods consist of adding information to a sequential language that eases parallel actors' extraction. With Open Multiprocessing (OpenMP [15]) for instance, a programmer can annotate C, C++ or Fortran code and indicate where he wants parallelism to be extracted. The CILK language [16] from the Massachusetts Institute of Technology is another example of such a technology adapted to C++. The compiler automates the mapping and transfers of parallel actors. OpenMP is already supported by the majority of the C, C++ and Fortran compilers. It was developed for Symmetric Multi-Processor (SMP) architectures and is likely to offer sub-optimal solutions for heterogeneous architectures. Other language extensions for parallelism that exist are less automated: Message Passing Interface (MPI), OpenCL [17] or the Multicore Association API [18] which offers APIs for manual multi-core message passing or buffer sharing.
- `Platform-based` methods actually refers to a single solution based on a Common Intermediate Code (CIC) [19], equivalent to a compiler Intermediate Representation (IR) with parallelism expressivity. It is intended to create an abstraction of the platform architecture to the compiler front-end that extracts parallelism.
- `Model of Computation (MoC)-based` design methods describe high-level application behavior with simple models. While compilation-based parallelization appeals to programmers because they can almost seamlessly reuse their huge amount of existing sequential code (generally in C or C++), MoC-based design methods bring several advantages compared with previous methods. They are aimed at expressing high-level algorithm descriptions into simple textual files or graph-based Graphical User Interfaces. MoC-based methods have three great advantages. Firstly, parallelism is naturally expressed by a graph. An extraction with a compilation algorithm will always under-performing on certain cases. The first step of Fig. 4.2 becomes straightforward. Secondly, certain rules of a model-

based design (such as schedulability) can be verified at compile-time and correct-by-construction high-level code may be generated. Thirdly, the transaction-level form of dataflow graphs can enable co-simulation of entities of different forms. This is why the Transaction-Level Modeling (TLM [20]) of SystemC is based on a dataflow MoC. Hardware-defined modules can be interconnected with software-defined modules.

As explained in Chap. 3, dataflow MoCs have been chosen to represent algorithms in the development of the rapid prototyping method and the prototyping framework PREESM.

4.3.2 PREESM Among Multi-Core Programming Tools

A brief overview on existing tools that support Data Flow programming is presented below.

4.3.2.1 Dataflow Interchange Format

Dataflow Interchange Format (DIF) [21] is a language for specifying dataflow models for DSP systems developed at the University of Maryland. DIF aims to provide an extensible repository of models, analysis, transformation, scheduling and code generation for the dataflow paradigm. The dataflow models currently supported are:

- Synchronous Dataflow.
- Homogeneous Synchronous Dataflow.
- Single Rate Dataflow.
- Cyclo-static Synchronous Dataflow.
- Parameterized Synchronous Dataflow.
- Multidimensional Synchronous Dataflow.
- Boolean Controlled Dataflow.
- Enable Invoke Dataflow.
- Core Functionnal Dataflow.

The DIF-to-C transformation [22] allows the generation of C code from an SDF specification. To achieve this, the DIF package (Fig. 4.3) provides an extensive repository of scheduling [Flat, Flattening, Hierarchical, APGAN (Algorithm for Pairwise Grouping of Adjacent Node)] and buffering techniques (Non shared, Circular, Buffer Sharing, Static, In-place Buffer Merging). The C code generated targets mainly DSP processors.

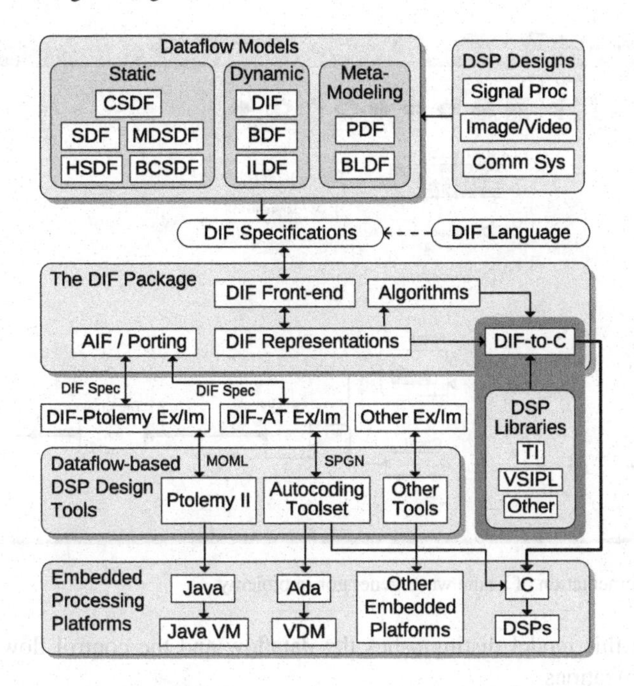

Fig. 4.3 DSP system design with DIF

4.3.2.2 LIDE

LIDE (the DSPCAD Lightweight Dataflow Environment) [23, 24] is a design environment aids the designer with the implementation of data flow based DSP applications. It consists of a library implementing semantic elements of the data flow paradigm such as actors, edges and schedulers. This tool when used with DICE (DSPCAD Integrative Command Line Environment) provides a full development and testing framework. Current version supports C language and HDL (verilog) and Java versions are being developped.

4.3.2.3 SDF3

SDF3 [25] is a tool for generating random Synchronous DataFlow Graphs (SDFGs), if desirable with certain guaranteed properties like strongly connectedness. It includes an extensive library of SDFG analysis and transformation algorithms as well as functionality to visualize them. This tool can create SDFG benchmarks that mimic DSP or multimedia applications. In addition, the tool supports the Scenario Aware Dataflow model [26] that allows parameterized production/consumption rates to be specified, and for which the actor execution time depends on the active scenario.

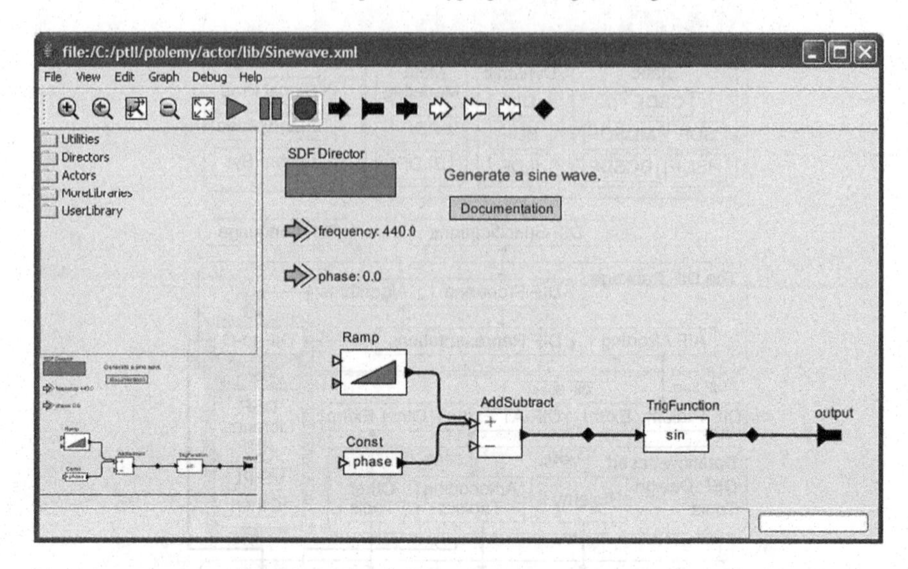

Fig. 4.4 Implementation of a sine wave generator in ptolemy

Furthermore this model distinguishes the dataflow and the control flow allowing further optimizations.

4.3.2.4 Ptolemy II

Ptolemy [27] is a tool developed at Berkeley for application modeling and simulation. An application may be described using a composition of actors called networks directed by a given computation model called the director. It supports a wide collection of computation models, from event-driven to the majority of data-flow models and process networks. Ptolemy II is the successor of Ptolemy classic and has been under development since 1996. In this tool, a specification can be executed and debugged using underlying java code. The tool is supported by a graphical environment called VIRGIL (Fig. 4.4), that allows graph specification and graphical application debugging. The models are stored in a concrete XML syntax called MoML.

4.3.2.5 StreamIt

StreamIt [28] is a programming language and a compilation infrastructure developed at the MIT, and is specifically engineered for modern streaming systems. It is designed to facilitate the programming of large streaming applications, as well as their efficient and effective mapping to a wide variety of target architectures, including commercial-off-the-shelf uniprocessors, multi-core architectures, and clusters of workstations. The specification is made using a textual representation. Each actor is declared as a

```
int->int loop matVectProd(int N){
        join roundrobin(1, N) ;
        add vectScalarProd();
        loop propagate();
        split duplicate ;
        enqueue 0.0;
}

int->int filter propagate(){
        prework push N{
                push(0.0);
        }
        work push 1 pop 1{
                push(pop());
        }
}

int->int loop vectScalarProd(){
        join roundrobin ;
        add mac();
        loop propagate();
        split duplicate ;
        enqueue 0.0;
}

int->int filter mac(void){
        work push 1 pop 3{
                a = pop();
                b = pop();
                c = pop();
                push(c * b + a);
        }
}
```

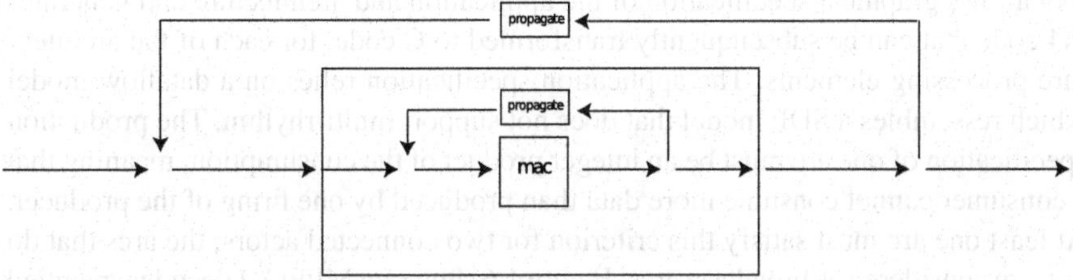

Fig. 4.5 StreamIt model of the matrix vector product example

filter. A Filter is an actor with exactly one input and one output port. It can declare an initialization function (init), a pre-work function and a work function. The init function is executed at the instantiation and does not consume/produce tokens. The pre-work function is the very first execution of the filter. The work function defines the behavior of the filter. Work and pre-work are characterized by push and pop properties that define the production and consumption of the filter (see Fig. 4.5). Filters can support dynamic rates. The network of actors is defined as a pipeline in which filters are added in a sequential order. A split-join section may be used; this is defined as a section of filters in which the input data is split and the output is the result of the join of several filters. The modelization language is supported by a command-line compiler that can be setup with different optimization options.

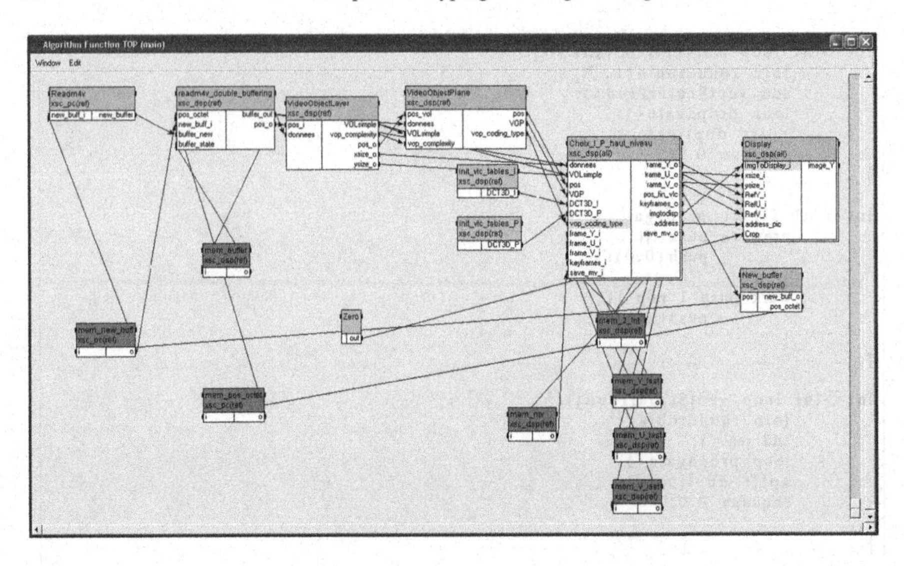

Fig. 4.6 MPEG-4 decoder model specified in SynDEx

4.3.2.6 SynDEx Data-Flow Model

SynDEx is a rapid prototyping tool developed at the INRIA Rocquencourt [29]. This tool allows graphical specification of the application and architecture and generates m4 code that can be subsenquently transformed to C code, for each of the architecture processing elements. The application specification relies on a dataflow model which ressembles a SDF model that does not support multi-rhythm. The production specification of one arc must be an integer product of the consumption, meaning that a consumer cannot consume more data than produced by one firing of the producer. At least one arc must satisfy this criterion for two connected actors; the arcs that do not, are considered as broadcast arcs. Figure 4.6 shows an MPEG-4 decoder specified in SynDEx tool.

4.3.2.7 Canals

Canals [30] is a data flow modeling language supported by a compiler infrastructure. This language and its compiler are being developed at the University of Turku in Finland by the Center for Reliable Software Technology. This language is based on the concept of nodes (kernel and networks) and links which connect these nodes together. A node or kernel is a computational element that takes a fixed amount of tokens on its input and a fixed or arbitrary number of element on its output. A kernel can also *look* at values on its input port without consuming it. Special kernels *scatter* and *gather* are used for data distribution and grouping with a specified behavior respectively.

In this language, the scheduling is separated from the network, and is specified by the user for the whole network. The scheduler element is responsible for planning the execution of kernels, in such a way that input data to the network is routed correctly through the network and eventually becomes output. The scheduler can accomplish correct routing by inspecting the arriving data, at run-time, and make routing decisions based on the contents of the data.

The Canals compiler is then able to provide a mapping of the top network onto a given architecture and generate code for each programmable component. The C, C++ code genrated is platform independent and the hardware specific resources and communication links are handled by a Hardware Abstraction Layer library for each component.

4.3.2.8 CAL Actor Language

RVC-CAL [31, 32] is a subset of the CAL Actor Language [33] used for defining the functionality of dataflow components called *actors* and so writing dataflow models.

The CAL Actor Language was first created as part of the Ptolemy project [34]. Actors were originally written in Java with a complex API, but ultimately this solution proved unnecessarily difficult to use in practice, and the CAL Actor Language was consequently designed. An actor defines input ports and output ports, the variables containing its internal state, and a number of transition rules called actions. Each actor executes by making discrete, atomic steps or transitions. In each step, exactly one action is chosen from its set of actions (according to the conditions associated with that action), and then executed. This execution consists of a combination of the following:

1. consumes input tokens,
2. produces output tokens,
3. modifies the state of the actor.

The profiled version of CAL Actor Language, called RVC-CAL, has been selected by the ISO/IEC standardization organization in the new MPEG Reconfigurable Video Coding (RVC) standard [32]. By constrast with CAL, RVC-CAL restricts the data types, the operators, and the features that may be used in actors. One driver for creating RVC-CAL is the difficulty of implementing a code generator that outputs correct and efficient code for CAL in its entirety. Both software code and hardware code fro RVC-CAL can be efficiently generated from CAL. However, the original expressiveness of CAL and the strong encapsulation features offered by its actor programming model have been preserved and provide a solid foundation for the compact and modular specification of actors in RVC-CAL.

Two tools support the CAL language. The OpenDF framework allows the network-level simulation of CAL actors and the generation of HDL description, while ORCC (Open RVC CAL Compiler) only supports the RVC-CAL subset but generates C, LLVM, and HDL code.

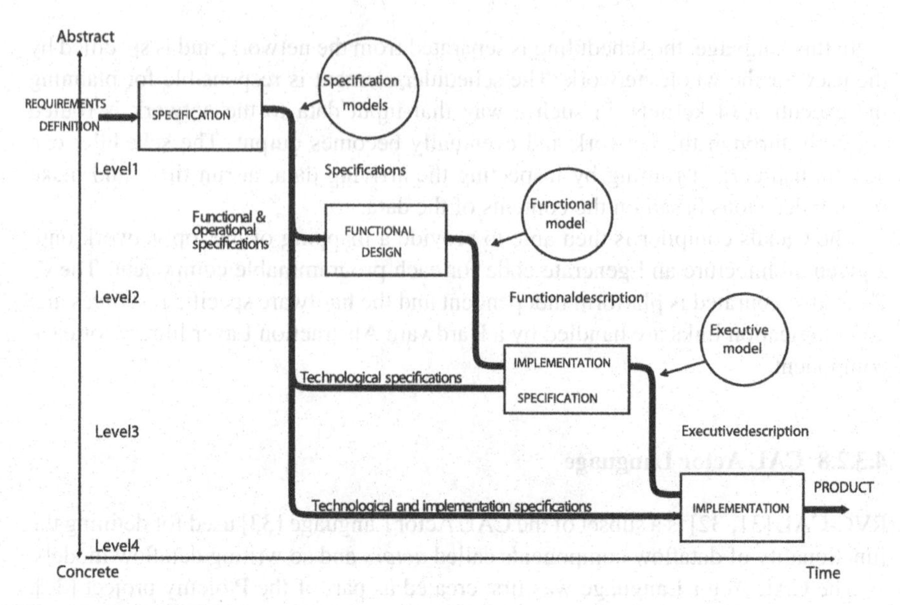

Fig. 4.7 The MCSE design cycle

4.3.2.9 MCSE Specification Model

MCSE [35] (Methode de Conception de Système Electronique) is a french acronym that stands for Electronic System Design Methodology. It was developed at the polytechnic school of Nantes (France) by Jean Paul Calvez. This methodology aims to provide the systems engineer with a work-flow for systems design. This work-flow act as a guide from the customer requirement to a valid prototype. This process goes through several design steps associated with a relevant model. The process steps (Fig. 4.7) are:

1. Specification
2. Functional Design
3. Implementation specification
4. Implementation

The model is refined at each step leading to an execution model that may be simulated and synthesized. The functional model is an application model that mixes both data-flow and event driven models (Fig. 4.8). The use of the shared variables produces a the model resembling a DDF model in that a variable can be accessed without consuming it. No formal verification of the model behavior is performed but metrics such as operator load and memory footprint can be extracted.

The design work-flow allows the validation of the designer choice at each step leading to an acceptable solution. The resulting automatic documentation of each step is advantageous for communication within the team and with the customer.

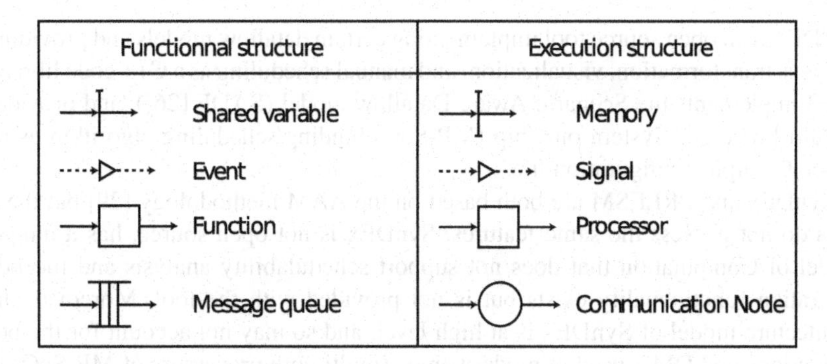

Fig. 4.8 The MCSE model components

4.3.2.10 Other Methods

Numerous solutions to partition algorithms onto multi-core architectures exist. For a homogeneous target architecture multi-core applications may be generated from C with additional information (OpenMP [15], CILK [16]). For heterogeneous architectures, languages such as OpenCL [17] and the Multicore Association Application Programming Interface (MCAPI [18]) define ways to express parallel properties of a code. However, these languages are not currently linked to efficient compilers and run-time environments. Moreover, compilers for such languages would have difficulty in extracting and solving the bottlenecks inherent to graph descriptions of the architecture and the algorithm.

The Poly-Mapper tool from PolyCore Software [36] offers functionalities similar to PREESM but, in contrast to PREESM, has manual mapping/scheduling. Ptolemy II [27] is a simulation tool that supports many computational models. However, it has no automatic mapping and its current code generation for embedded systems is focused on single-core targets. Another family of frameworks which exists for dataflow based programming is based on CAL [37] language and includes OpenDF [38]. OpenDF employs a more dynamic model than PREESM but its related code generation does not currently support multi-core embedded systems.

Closer in principle to PREESM are the Model Integrated Computing (MIC [39]), the Open Tool Integration Environment (OTIE [40]), the Synchronous Distributed Executives (SynDEx [41]), the Dataflow Interchange Format (DIF [21]), and SDF for Free (SDF3 [42]). Neither MIC nor OTIE can be downloaded from internet. According to the literature, MIC focuses on the transformation between algorithm domain-specific models and meta-models while OTIE defines a single system description that can be used during the whole signal processing design cycle.

DIF is designed as an extensible repository of representation, analysis, transformation and scheduling of several dataflow languages. DIF is a Java library which allows the user to design graph specification to C code generation using the DIF language. However, the IBSDF model (Sect. 3.5.2) as used in PREESM is not available in DIF.

SDF3 is an open source tool implementing certain dataflow models and providing analysis, transformation, visualization, and manual scheduling as a C++ code library. SDF3 implements the Scenario Aware Dataflow model (SADF [26]), and provides a Multi-Processor System-on-Chip (MP-SoC) binding/scheduling algorithm as an MP-SoC output configuration file.

SynDEx and PREESM are both based on the AAM methodology [29] but these tools do not possess the same features. SynDEx is not open source, has a unique Model of Computation that does not support schedulability analysis and thecode generation functionnality exists but is not provided with the tool. Moreover, the architecture model of SynDEx is at high level, and so may not account for the bus contentions and DMA used in modern chips (multi-core processors of MP-SoC) in mapping/scheduling.

The features that differentiate PREESM from the similar tools are:

- The tool is open source and accessible online,
- The algorithm description is based on a single well-known and predictable model of computation,
- The scheduling is totally automatic,
- The functional code for heterogeneous multi-core embedded systems can be generated automatically,
- The IBSDF algorithm model provides an efficient hierarchical encapsulation thus simplifying the scheduling [43].

The central feature of the rapid prototyping method used in both this study and the PREESM tool is the multi-core scheduler. The next section presents the multi-core scheduling techniques proposed in this study, and which are based on algorithms well documented in the literature.

4.4 Multi-Core Scheduling

4.4.1 Multi-Core Scheduling Strategies

In [44], Lee develops a taxonomy of scheduling algorithms that distinguishes run-time and compile-time scheduling steps. The choice of the scheduling steps executed at compile-time and the ones executed at run-time is called a scheduling strategy. The taxonomy of scheduling strategies is summarized in Table 4.2. These strategies are displayed on the left-hand side column and are ordered from the most dynamic to the most static.

The principle in [44] is that a maximal number of operation should be executed at compile time because it reduces the execution overhead due to run-time scheduling. For a given algorithm, the lowest possible scheduling strategy in the table that enables suited schedules should thus be used.

Table 4.2 Scheduling strategies

	Assignment	Ordering	Timing
Fully dynamic	Run	Run	Run
Static-assignment	Compile	Run	Run
Self-timed	Compile	Compile	Run
Fully static	Compile	Compile	Compile

The fully static strategy is based on the idea that the exact actor relations and execution times are known at compile-time. When this is the case, the operators do not need to be synchronized at run-time, they only need to produce and consume their data at the right moment in the cycle. However, the modern DSP architectures with caches and RTOS that manipulate threads do not qualify for this type of strategy. The self-timed strategy which still assigns and orders at compile-time but synchronizes the operators is well suited to execute a SDF graph on DSP architecture. This strategy has been the subject of much study in [45] and will be used in Chap. 8. The static assignment strategy consists of only selecting the assignment at compile time and delegating the task to order and fire actors to a RTOS on each core. This method is quite natural to understand because the RTOS on each core can exploit decades of research on efficient run-time single-core thread scheduling. Finally, the fully dynamic strategy makes all choices at run-time. Being very costly, fully dynamic schedulers usually employ very simple and sub-optimal heuristics. In Sect. 8.3, a technique implementing a fully dynamic strategy and named adaptive scheduling is introduced. Adaptive scheduling schedules efficiently the highly variable algorithms of LTE.

The scheduling techniques for multi-core code generation most applicable for LTE physical layer algorithms are the self-timed and adaptive scheduling ones. [44] claims that self-timed techniques are the most attractive software scheduling techniques in terms of tradeoff between scheduling overhead and run-time performance but the highly variable behavior of the LTE uplink and downlink algorithms compels the system designer to develop adaptive technique(Chap. 8).

4.4.2 Scheduling an Application Under Constraints

Using directed acyclic dataflow graphs as inputs, many studies have focused on the optimization of an application behavior on multi-core architectures. The main asset of these studies is the focus they can give on the middle-grain behavior of an algorithm. They naturally ignore the details of the instructions execution and deal with the correct partitioning of data-driven applications. Concurrent system scheduling is a very old problem. It has been extensively used, for example, to organize team projects or distributing tasks in factories.

A model of the execution must be defined to simulate it while making assignment and ordering choices. In this study, an actor or a transfer fires as soon as its input data

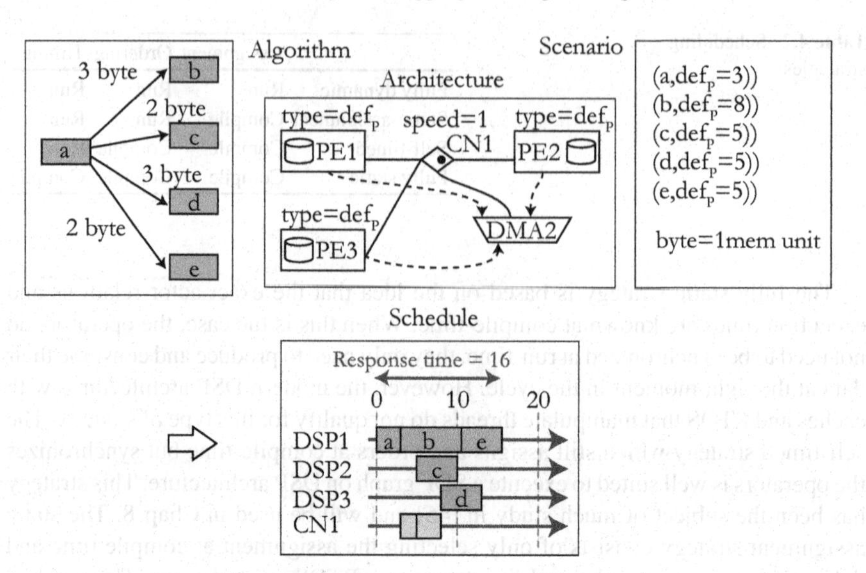

Fig. 4.9 Example of a schedule Gantt chart

is available. This execution scheme corresponds to a self-timed schedule (Sect. 4.4.1) [45]. Other execution schemes can be defined, for instance triggering actors only at periodic clock ticks. A self-timed multi-core schedule execution can be represented by a Gantt chart such as the one in Fig. 4.9. This schedule results from the scheduling of the given algorithm, architecture and scenario.

The scenario will be detailed in Chap. 6 and the architecture model in Chap. 5. The next section explains scheduling techniques.

Multi-core scheduling under constraints is a complex operation. This problem has been proven to be in the NP-complete (Non-deterministic Polynomial-time-complete) complexity class [46]. Many "useful" problems are NP-complete and [47] references many of them. The properties of NP-complete problems are that:

1. the verification that a possible solution of the problem is valid can be computed in polynomial time. In the scheduling case, verifying that a schedule is valid can be done in polynomial time [48].
2. an NP-complete problem can be "converted" into any other NP-complete problem in polynomial time. This means that solving one NP-complete problem in polynomial time would result in solving them all in polynomial time,
3. unfortunately, no polynomial time algorithm for NP-complete problems is known and it is likely that none exists.

Proving that a problem is NP-complete consists in finding a polynomial transformation from the considered problem to a known NP-complete problem. An example of such a polynomial transformation is given in [45] p.49. The multi-core scheduling problem is NP-complete, so heuristics must be found that offer a tradeoff between computational time and optimality of results.

4.4.3 Existing Work on Scheduling Heuristics

The literature on multi-core scheduling is extensive. In [46], Brucker gives a large overview of scheduling algorithms in many domains.

Assignment, ordering and timing can be executed either at compile-time or at run-time, defining a scheduling strategy. In his Ph.D. thesis [49], Kwok gives an overview of the existing work on multi-core compile-time scheduling and puts forward three low complexity algorithms that he tested over random graphs with high complexity (in the order of 1,000 to 10,000 actors). In this Ph.D., the objective function is always the latency of one graph iteration.

Multi-core scheduling heuristics can be divided into three categories depending on their target architecture:

1. `Unbounded Number Of Clusters` (UNC) heuristics consider architectures with infinite parallelism, totally connected and without transfer contention.
2. `Bounded Number of Processors` (BNP) heuristics consider architectures with a limited number of operators, totally connected and without transfer contention.
3. `Arbitrary Processor Network` (APN) heuristics consider architectures with a limited number of operators, not totally connected and with transfer contention.

From the first UNC heuristic by Hu in 1961 [50], many heuristics have been designed to solve the BNP and then the APN generic problems. The architecture model of the APN is the most precise one. This category needs to be addressed because modeled heterogeneous architectures are not totally connected.

Most of the heuristics enter in the BNP `static list scheduling` category. A static list scheduling takes as inputs:

- A DAG $G = (V, E)$,
- an execution time for each vertex v in V which is the same on every operator because each is homogeneous,
- a transfer time for each edge e in E which is the same between any two operators because they are totally connected with perfectly parallel media,
- a number of operators P.

For each actor in the DAG (Sect. 3.2), timings can be computed that feed the scheduling process. DAG timings are displayed in Fig. 4.10; they include:

1. The `t-level` or `ASAP` (As Soon As Possible) starting time of an actor v is the time of the longest path from an entry actor to v. It represents the earliest execution time of v if no assignment and ordering constraint delays v.
2. The `b-level` of v is the time of the longest path from v to an exit actor (including the duration of v).

The `Critical Path` (CP) is the longest path between an entry actor and an exit actor in the DAG (Sect. 3.2.3) considering actor and edge durations. In Fig. 4.10,

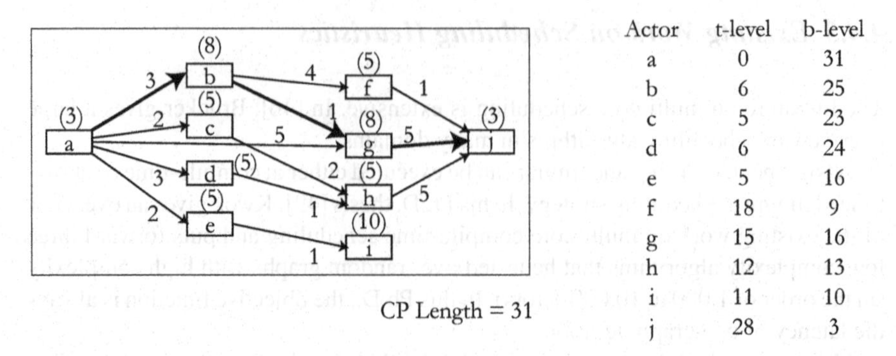

Actor	t-level	b-level
a	0	31
b	6	25
c	5	23
d	6	24
e	5	16
f	18	9
g	15	16
h	12	13
i	11	10
j	28	3

Fig. 4.10 T-level and B-level of a timed directed acyclic graph

the critical path is shown with larger lines. Timing a whole graph has a complexity $O(|E|+|V|)$ because each actor and edge must be scanned once (forward for t-level and backward for b-level).

A static list scheduling heuristic consists of three steps:

1. `Find a DAG topological order`. A topological order is a linear ordering of the vertices of the DAG $G = (V, E)$ such that if there is a directed edge between two actors u and v, u appears before v [51]. In [52], Mu gives an algorithm to find an arbitrary topological order with a complexity $O(|V| + |E|)$, which is the minimal possible complexity [53]. A directed graph is acyclic if and only if a topological order of its actors exists.
2. `Construct an actor list` from the topological order. This step is straightforward but necessitates a list ordering of complexity at least $O(|V| \cdot log(|V|))$ [53].
3. `Assign and order actors` in order of the actor list without questioning the previous choices. The usual method to assign actors is to choose the operator that offers the earliest start time and add the actor at the end of its schedule. Kwok gives such an algorithm in [49] with a complexity $O(P \cdot (|E| + |V|))$.

Static list scheduling heuristics are "greedy" algorithms because they make locally optimal choice but never consider these choices within the context of the whole system. Kwok gives an algorithm in $O(|E| + |V|)$ to find an optimized actor list named `CPN-dominant list` based on the DAG critical path. Actors are divided into three sets:

1. `Critical Path Nodes` (CPN) are actors that belong to the critical path. To ensure a small latency, they must be allocated as soon as possible.
2. `In-Branch Nodes` (IBN) are actors that are not CPN. For each IBN v, a path exists between v and a CPN node or between a CPN node and v. This property means that scheduling choices of IBN nodes are likely to affect the CPN choices. They have a medium priority while mapping.

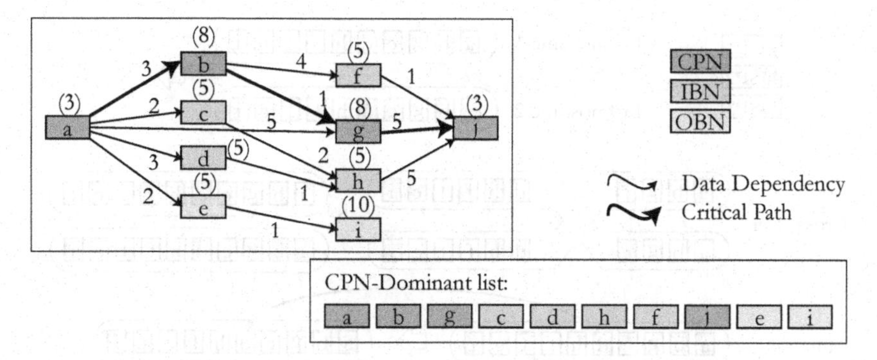

Fig. 4.11 Construction of an actor list in topological order

3. `Out-Branch Nodes` (OBN) are the actors that are neither CPN nor IBN. Their ordered execution on operators can also affect the execution latency but with lower probability than CPN and IBN. They have a low priority while mapping.

In the CPN-Dominant list, CPN nodes are inserted as soon as possible and OBN as late as possible while respecting the topological order. Figure 4.11 shows a CPN-Dominant list and the types (CPN, IBN or OBN) of the actors. The global complexity of the static list scheduling using a CPN-dominant list is $O(|V| \cdot log(|V|) + P \cdot (|E| + |V|))$.

Dynamic list scheduling is another family of scheduling heuristics in which the order of the actors in the priority list is recomputed during the ordering process. [54] advises against dynamic list scheduling heuristics because their slightly improved results compared to static list scheduling do not justify their higher complexity.

The FAST (Fast Assignment and Scheduling of Tasks) algorithm enters in the `neighborhood search` category. This algorithm starts from an initial solution, which it then refines, while testing local modifications. The FAST algorithm is presented in [49, 55]. The FAST algorithm can run for an arbitrary length of time defined by user to refine the original solution. It consists of macro-steps in which CPN nodes are randomly selected and moved to another operator. One macro-step contains many micro-steps in which IBN and OBN nodes are randomly picked and moved to another operator. At each micro-step, the schedule length is evaluated and the best schedule is kept as a reference for the next macro-step. There is hidden complexity in the micro-step because the t-level of all the actors affected by the new assignment of an actor v must be recomputed to evaluate the schedule length. The affected actors include all the successors of v as well as all its followers in the schedules of its origin and target operators and their successors. Generally, these vertices represent a substantial part of the DAG graph. One micro step is thus always at least $O(|E| + |V|)$.

A Parallel FAST algorithm (or PFAST) is described in [49] and is an algorithm which can multi-thread the FAST execution. The CPN-dominant list is divided into sub-lists in topological order. Each thread processes the FAST on one sub-list, the

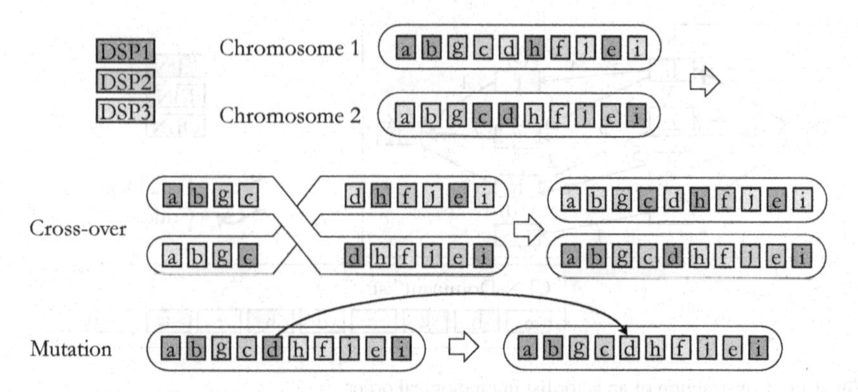

Fig. 4.12 Genetic algorithm atomic operations: mutation and cross-over

results are grouped periodically and the best scheduling is kept as a starting point for the next FAST iteration. The PFAST heuristic is implemented in PREESM.

A great advantage of the FAST algorithm is that it generates many schedules over time. A "population" of schedules can be generated by storing the *N* found schedules that had the lowest latency. In the population, each schedule ordered following the CPN-Dominant list is called a chromosome and each actor assignment is called a gene. For this initial population, a genetic algorithm can be launched on assignments and recursively the two operations displayed in Fig. 4.12:

- A `mutation` randomly changes the assignment of one randomly selected actor (gene) in the schedule (chromosome).
- A one-point `cross-over` mixes two chromosomes, exchanging their assignments from the beginning to the crossover point.

After each step, which includes one mutation and one cross-over, the schedule latency is evaluated and the best schedule is kept.

All the above techniques are BNP scheduling heuristics and do not take into account inter-processor transfer ordering or contention between transfers sharing the same resource. [51, 52] address this problem known as `edge scheduling`. The edge scheduling problem under constraints is NP complete [56]. In Chap. 6, heuristics for edge scheduling are developed within a scheduling framework. PREESM uses modified versions of the list, FAST and genetic algorithms described above. These modified algorithms are discussed in Chap. 6.

4.5 Generating Multi-Core Executable Code

The scheduling strategy (Sect. 4.4.1) greatly influences how code is generated from a scheduled graph. Two main choices exist: generating a static execution when no OS is necessary or to rely on an OS capable of adaptive scheduling. The term "adaptive" is used instead of "dynamic" to avoid confusion with dynamic list scheduling (Sect. 4.4.3).

4.5.1 Static Multi-Core Code Execution

For applications that do not vary greatly from computed behavior, i.e. close estimates of its actor timings are computable, the code may be generated following the self-timed scheduling strategy (Sect. 4.4.1). The Algorithm Architecture Matching method (AAM, previously AAA [41]) is a method which generates self-timed coordination code from a quasi-static schedule of a dataflow graph. The MoC used is specific to the method and can contain conditions. Linear executions on each operator are interconnected with data transfers and synchronized with semaphores.

In [57], Bhattacharyya et al. present a self-timed code execution technique with reduced synchronization costs for the SDF MoC. The optimization removes the synchronization point and still protects data coherency when successive iterations of the graph overlap. Synchronizations are the overhead of a self-timed execution compared to a fully-static one. Synchronizations between the linear executions on each operator must be minimized for the code execution performances to match the simulation Gantt chart. Additionally, the buffer sizes are bounded to enable execution with limited FIFO sizes. Ideas in [57] are extended in [45].

In [56, 58], the authors add to the minimization of the inter-core synchronization the idea of communication ordering. This idea, equivalent to edge scheduling (Sect. 4.4.3), can greatly reduce the latency of the graph execution.

In Sect. 8.2, a code generation technique is developed and applied to the Random Access Procedure part of the LTE physical layer processing.

4.5.2 Managing Application Variations

The study of LTE will show that some algorithms are highly variable over time. In [59], Balarin et al. present an overview of the scheduling data problem and control-dominated signal processing algorithms. The lack of good models for scheduling adaptively algorithms when their properties vary over time is noted. [60] develops quasi-static scheduling techniques which schedule dynamic graphs at compile time. Quasi-static scheduling seeks an algorithm that enables good system parallelism in all cases of algorithm execution.

Studies exist that schedule variable algorithms at run-time. The Canals language [30] constitutes an interesting hierarchical dataflow approach to such run-time scheduling that appears similar to the hierarchical models in the Ptolemy II rapid prototyping tool, with each block having a scheduler [61] and a high-level scheduler invoking the schedulers of its constituent blocks. The Canals language focuses on image processing algorithms that can be modeled by a flow-shop problem [46], i.e. several independent jobs each composed of sequential actors. Another dataflow run-time scheduling method that focuses on flow-shop problems is developed in [62].

Section 8.3 presents an adaptive scheduler that uses a simplified version of the list scheduling algorithm in Sect. 4.4.3 and solves the LTE uplink scheduling problem in real-time on embedded platforms. In this scheduler, dataflow graphs are used as an Intermediate Representation for a Just-In-Time multi-core scheduling process.

4.6 Methods for the LTE Study

The previous chapters introduced three different domains necessary to understand multi-core and dataflow-based LTE rapid prototyping. The features of the LTE standard were detailed in Chap. 2. Chapter 3 explained the concept of dataflow Model of Computation that will be used to model LTE. Finally, existing work on rapid prototyping and multi-core programming were developed in this chapter.

In the next chapters, methods are explained that enhance the process of rapid prototyping: a new architecture model in Chap. 5, advanced rapid prototyping techniques in Chap. 6, LTE dataflow models in Chap. 7 and LTE multi-core code generation in Chap. 8.

References

1. Davis AL, Stotzer EJ, Tatge RE, Ward AS (1996) Approaching peak performance with compiled code on a VLIW DSP. Proceedings of ICSPAT Fall
2. Lee E (2006) The problem with threads. Computer 39(5):33–42
3. Lee EA, Parks TM (1995) Dataflow process networks. Proc IEEE 83(5):773–801
4. International Technology Roadmap for Semiconductors (2009) I.T.R.: Design. www.itrs.net
5. Karam LJ, AlKamal I, Gatherer A, Frantz GA, Anderson DV, Evans BL (2009) Trends in multicore DSP platforms. IEEE Signal Process Mag 26(6):38–49
6. Heath S (1995) Microprocessor architectures RISC, CISC and DSP. Butterworth-Heinemann Ltd., Oxford
7. Tensilica. http://www.tensilica.com/
8. Flynn MJ (1972) Some computer organizations and their effectiveness. IEEE Trans Comput 100:21
9. Blazewicz J, Ecker K, Plateau B, Trystram D (2000) Handbook on parallel and distributed processing. Springer, Berlin
10. Open SystemC initiative web site. http://www.systemc.org/home/
11. Feiler PH, Gluch DP, Hudak JJ (2006) The architecture analysis and design language (AADL): an introduction. Technical report
12. Graham RL (1966) Bounds for certain multiprocessing anomalies. Bell Syst Tech J 45(9):1563–1581
13. Park H, Oh H, Ha S (2009) Multiprocessor SoC design methods and tools. IEEE Signal Process Mag 26(6):72–79
14. Ceng J, Castrillon J, Sheng W, Scharwachter H, Leupers R, Ascheid G, Meyr H, Isshiki T, Kunieda H (2008) MAPS: an integrated framework for MPSoC application parallelization. In: Proceedings of the 45th annual conference on design automation, pp. 754–759
15. OpenMP. http://openmp.org/wp/
16. Blumofe RD, Joerg CF, Kuszmaul BC, Leiserson CE, Randall KH, Zhou Y (1995) Cilk: an efficient multithreaded runtime system. In: Proceedings of the fifth ACM SIGPLAN symposium on principles and practice of parallel programming
17. OpenCL. http://www.khronos.org/opencl/
18. The Multicore Association. http://www.multicore-association.org/home.php
19. Kwon S, Kim Y, Jeun WC, Ha S, Paek Y (2008) A retargetable parallel-programming framework for MPSoC. ACM Trans Des Autom Electron Syst 13(3):39
20. Ghenassia F (2006) Transaction-level modeling with systemC: TLM concepts and applications for embedded systems. Springer-Verlag New York, Inc. http://portal.acm.org/citation.cfm?id=1213675

21. Hsu CJ, Keceli F, Ko MY, Shahparnia S, Bhattacharyya SS (2004) Dif: an interchange format for dataflow-based design tools. http://citeseerx.ist.psu.edu/viewdoc/summary? doi:10.1.1.76.6859
22. Hsu, C.J., Ko, M.Y., Bhattacharyya, S.S.: Software synthesis from the dataflow interchange format. In: SCOPES '05: Proceedings of the 2005 workshop on Software and compilers for embedded systems, pp. 37–49. ACM, New York, NY, USA (2005). http://doi.acm.org/10.1145/1140389.1140394
23. Shen C, Plishker W, Wu H, Bhattacharyya SS (2010) A lightweight dataflow approach for design and implementation of SDR systems. In: Proceedings of the wireless innovation conference and product exposition, Washington DC, USA, pp. 640–645
24. Shen C, Wang L, Cho I, Kim S, Won S, Plishker W, Bhattacharyya SS (2011) The DSP-CAD lightweight dataflow environment: introduction to LIDE version 0.1. Technical report UMIACS-TR-2011-17, Institute for Advanced Computer Studies, University of Maryland at College Park http://hdl.handle.net/1903/12147
25. Stuijk S, Geilen M, Basten T (2006) SDF^3: SDF For Free. In: Application of concurrency to system design, 6th international conference, ACSD 2006, proceedings. IEEE Computer Society Press, Los Alamitos, CA, USA pp. 276–278. http://www.es.ele.tue.nl/sdf3
26. Theelen B (2007) A performance analysis tool for Scenario-Aware streaming applications. In: Fourth international conference on the quantitative evaluation of systems, QEST 2007, pp. 269–270. doi:10.1109/QEST.2007.7
27. Lee E (2001) Overview of the ptolemy project. Technical memorandum UCB/ERL M01/11, University of California at Berkeley
28. Thies, W., Karczmarek, M., Gordon, M., Maze, D.Z., Wong, J., Hoffman, H., Brown, M., Amarasinghe, S.: Streamit: A compiler for streaming applications. Technical Report MIT/LCS Technical Memo LCS-TM-622, Massachusetts Institute of Technology, Cambridge, MA (2001). http://groups.csail.mit.edu/commit/papers/01/StreamIt-TM-622.pdf
29. Grandpierre T, Sorel Y (2003) From algorithm and architecture specifications to automatic generation of distributed real-time executives: a seamless flow of graphs transformations. In: MEMOCODE '03, pp. 123–132
30. Dahlin A, Ersfolk J, Yang G, Habli H, Lilius J (2009) The canals language and its compiler. In: Proceedings of th 12th international workshop on software and compilers for embedded systems, pp. 43–52
31. Bhattacharyya SS, Eker J, Janneck JW, Lucarz C, Mattavelli M, Raulet M (2009) Overview of the MPEG Reconfigurable Video Coding Framework. Springer journal of Signal Processing Systems, Special Issue on Reconfigurable Video Coding, 2009
32. ISO/IEC FDIS 23001–4: MPEG systems technologies—Part 4: Codec Configuration Representation (2009)
33. Eker J, Janneck J (2003) CAL Language Report. Technical Report, ERL Technical Memo UCB/ERL M03/48, University of California at Berkeley, 2003
34. Eker J, Janneck J, Lee E, Liu J, Liu X, Ludvig J, Neuendor S, Sonia E, Yuhong S (2003) Taming heterogeneity—the Ptolemy approach. In. Proceedings of the IEEE, vol. 91, 2003
35. Calvez JP, Isidoro D (1994) A codesign experience with the mcse methodology. In: CODES '94: Proceedings of the 3rd international workshop on Hardware/software co-design, pp. 140–147. IEEE Computer Society Press, Los Alamitos, CA, USA, 1994
36. PolyCore Software Poly-Mapper tool. http://www.polycoresoftware.com/products3.php
37. Eker J, Janneck JW (2003) CAL Language Report. Technical report, ERL Technical Memo UCB/ERL M03/48, University of California at Berkeley, 2003
38. Bhattacharyya S, Brebner G, Eker J, Janneck J, Mattavelli M, von Platen C, Raulet M (2008) OpenDF—a dataflow toolset for reconfigurable hardware and multicore systems. SIGARCH Comput. Archit, News
39. Karsai G, Sztipanovits J, Ledeczi A, Bapty T (2003) Model-integrated development of embedded software. Proc IEEE 91(1):145–164. doi:10.1109/JPROC.2002.805824
40. Belanovic P (2006) An open tool integration environment for efficient design of embedded systems in wireless communications. Ph.D. thesis, Technischen Universitat Wien

41. Grandpierre T, Lavarenne C, Sorel Y (1999) Optimized rapid prototyping for real-time embedded heterogeneous multiprocessors. In: (CODES '99) Proceedings of the Seventh International Workshop on, Hardware/Software Codesign, pp. 74–78, 1999
42. Stuijk S (2007) Predictable mapping of streaming applications on multiprocessors. Ph.D. thesis, Technische Universiteit Eindhoven
43. Piat J, Bhattacharyya SS, Pelcat M, Raulet M (2009) Multi-core code generation from interface based hierarchy. DASIP 2009
44. Lee EA (1989) Scheduling strategies for multiprocessor real-time DSP. In: IEEE global telecommunications conference and exhibition. Communications Technology for the 1990s and Beyond, 1989
45. Sriram S, Bhattacharyya SS (2009) Embedded multiprocessors: scheduling and synchronization, 2nd edn. CRC press, Boca Raton
46. Brucker P (2007) Scheduling algorithms. Springer, New York
47. Garey MR, Johnson DS (1990) Computers and intractability: a guide to the theory of NP-Completeness. W. H. Freeman & Co. (1990). http://portal.acm.org/citation.cfm?id=574848
48. Cormen TH, Leiserson CE, Rivest RL, Stein C (2001) Introduction to algorithms, 2nd edn. The MIT Press, Cambridge
49. Kwok Y (1997) High-performance algorithms of compile-time scheduling of parallel processors. Ph.D. thesis, Hong Kong University of Science and Technology
50. Hu TC (1961) Parallel sequencing and assembly line problems. Oper Res 9(6):841–848
51. Sinnen O (2007) Task scheduling for parallel systems (Wiley Series on Parallel and Distributed Computing). Wiley-Interscience, New York
52. Mu P (2009) Rapid prototyping methodology for parallel embedded systems. Ph.D. thesis, INSA Rennes
53. Radulescu A, van Gemund AJ (1999) On the complexity of list scheduling algorithms for distributed-memory systems. In: Proceedings of the 13th international conference on Supercomputing, pp. 68–75
54. Janecek TH (2003) Static vs. dynamic List-Scheduling performance comparison. Acta Polytech 43(6):23–28
55. Kwok YK, Ahmad I (1999) FASTEST: a practical low-complexity algorithm for compile-timeassignment of parallel programs to multiprocessors. IEEE Trans Parallel Distributed Syst 10(2):147–159
56. Khandelia M, Bambha NK, Bhattacharyya SS (2006) Contention-conscious transaction ordering in multiprocessors DSP systems. In: IEEE Transactions on signal processing, 2006
57. Bhattacharyya SS, Sriram S, Lee EA (1997) Optimizing synchronization in multiprocessor DSP systems. IEEE Trans Signal Process 45(6):1605–1618
58. Bambha N, Kianzad V, Khandelia M, Bhattacharyya SS (2002) Intermediate representations for design automation of multiprocessor DSP systems. Des Autom Embed. Syst 7(4):307–323
59. Balarin F, Lavagno L, Murthy P, Sangiovanni-vincentelli A (1998) Scheduling for embedded real-time systems. IEEE Des Test Comput 15(1):71–82
60. Ha S, Lee EA (1997) Compile-time scheduling of dynamic constructs in dataflow program graphs. IEEE Trans Comput 46:768–778
61. Buck J, Ha S, Lee EA, Messerschmitt DG (1994) Ptolemy: A framework for simulating and prototyping heterogeneous systems. Int J Comput Simul 4(2):155–182
62. Boutellier J, Bhattacharyya SS, Silven O (2009) A low-overhead scheduling methodology for fine-grained acceleration of signal processing systems. J Signal Process Syst 57(2):121–122

Chapter 5
A System-Level Architecture Model

5.1 Introduction

For the LTE physical layer to be properly prototyped, the target hardware architectures need to be specified at system-level, using a simple model focusing on architectural limitations. The System-Level Architecture Model (S-LAM), which enables such specifications, is presented in Sect. 5.2. Sections 5.2.4 and 5.3 explain how to compute routes between operators from an S-LAM specification and Sect. 5.4 shows how transfers on these routes are simulated. Finally, the role of the S-LAM model in the rapid prototyping process is discussed in Sect. 5.5.

5.1.1 Target Architectures

It is Multi-Processor System-on-Chip (MPSoC) and boards of interconnected MPSoCs which are the architectures of interest and will be modeled. Modeling the system-level behavior is intended to be highly efficient, and employs a careful trade-off between simulation accuracy and result precision. As the desired functionalities are generally software generation and hardware system-level simulation, the majority of the architecture details (local caches, core micro-architectures...) are hidden from system-level study by the compiler and libraries.

The hardware boards available for this LTE study are:

- One board containing two Texas Instruments TMS320TCI6488 processors (Fig. 5.1).
- One board containing one Texas Instruments TMS320TCI6486 processor (Fig. 5.3).

Both boards are based on DSP cores of type c64x+ [1]. A c64x+ core is a VLIW core that can execute up to 8 instructions per clock cycle. Instructions are 32- or 16-bit, implying an instruction word of up to 256 bits. The instructions themselves

M. Pelcat et al., *Physical Layer Multi-Core Prototyping*,
Lecture Notes in Electrical Engineering 171, DOI: 10.1007/978-1-4471-4210-2_5,

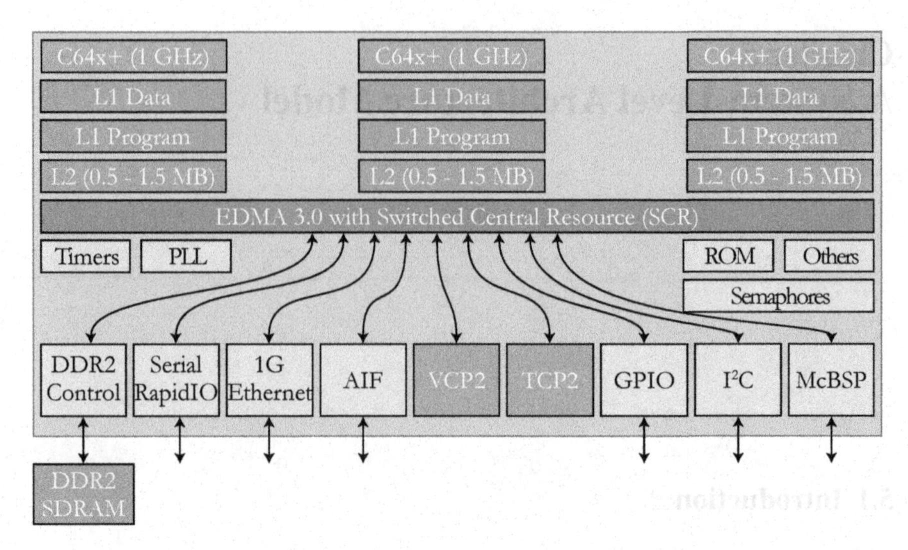

Fig. 5.1 TMS320TCI6488 functional block diagram

are SIMD: one c64x+ core embeds two multiply units, each one with the capacity of processing four 16- × 16-bit Multiply-Accumulates (MAC) per clock cycle. Each core can thus execute eight MAC per cycle.

Figure 5.1 shows a block diagram of the Texas Instruments TMS320TCI6488 processor [2]. For simplicity, this processor will be called tci6488. Each of the three cores of a tci6488 is a c64x+ core clocked at 1 GHz. The cores of a tci6488 can thus execute up to 24 GMAC per second (Giga Multiply-Accumulates per second) for a power consumption of several Watts.

No internal memory is shared by the cores. Each core has a 32 kB Level 1 program memory and a 32 kB Level 1 data memory. A pool of 3 MB of Level 2 memory can be distributed among the three cores in one of two configurations: 1/1/1 or 0.5/1/1.5 MB. The cores communicate with each other and with coprocessors and peripherals using an Enhanced Direct Memory Access named EDMA3. The EDMA3 is the only entity in the chip with access to the entire memory map. The processor uses a DDR2 interface to connect to a Double Data Rate external RAM.

The tci6488 has many embedded peripherals and coprocessors. The Ethernet Media Access Controller (EMAC) offers a Gigabit Ethernet access while the Antenna Interface (AIF) provides six serial links for retrieving digital data from the antenna ports. Each link has a data rate of up to 3.072 Gbps and is compliant with OBSAI and CPRI standards. A standardized serial RapidIO interface can be used to connect other devices with data rates up to 3.125 Gbps. The tci6488 has two coprocessors: the TCP2 and the VCP2 accelerate the turbo decoding and Viterbi decoding of a bit stream respectively (Chap. 2). Cores and peripherals are interconnected by the Switched Central Resource (SCR), which is an implementation of a switch. A switch is a hardware component that connects several components with ideally no contention.

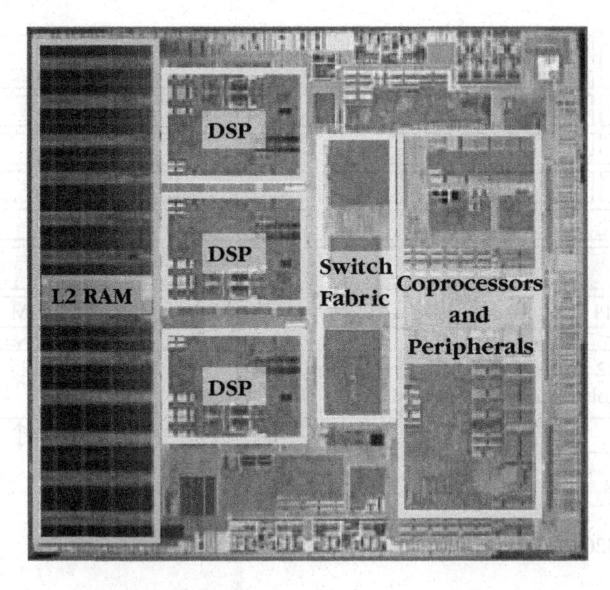

Fig. 5.2 Picture of a TMS320TCI6488 die

Sinnen gives three examples of switch implementations: the crossbar, the multistage network and the tree network ([3], p.18). Each offers a different tradeoff between hardware cost and contention management.

Figure 5.2 displays a photograph of a tci6488 die. It is fabricated using a 65 nm process technology. The internal memory, the three DSP cores, the Switched Central Resource, the coprocessors and peripherals are all clearly marked on the figure. It may also been seen that each c64x+ core represents only 8.5 % of the chip surface area despite the fact that it is a state-of-the-art high performance DSP core. Meanwhile, the 3 MB of internal memory represent one third of the chip surface area. It may be concluded that embedding several cores in a single DSP has significantly less impact on the chip size than providing extended on-chip memory. However, interconnecting cores and coprocessors is not a negligible operation: 11.5 % of the surface area is dedicated to the switch fabric interconnecting the elements.

Figure 5.3 is a block diagram of the internal structure of a TMS320TCI6486 processor. For simplicity, this processor will be referred to as tci6486. This processor has six embedded c64x+ cores clocked at 500 MHz. Like the tci6488, it can execute up to 24 GMAC per second. The six cores share a Level 2 Memory of 768 kB, with each locally containing 608 kB of Level 2 memory. Like the tci6488 cores, each core has a 32 kB Level 1 program memory and a 32 kB Level 1 data memory. The tci6486 has, among other peripherals, a serial RapidIO link, Gigabit Ethernet port and a controller of external DDR2 memory.

Previous paragraphs have focused on recent high performance DSP architectures targeted for base station applications. However, a recent LTE study details a new multi-core DSP architecture, which is not yet commercially named, and was

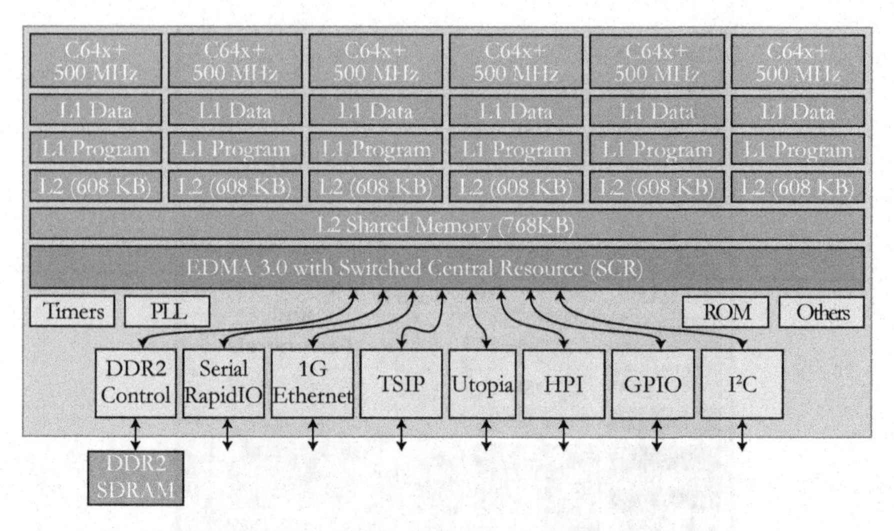

Fig. 5.3 TMS320TCI6486 functional block diagram

introduced in [4]. This architecture is specifically optimized for the LTE application. It has a clock speed of up to 1.2 GHz and a total capability of up to 256 GMAC per second. The new DSP cores of this chip are backward compatible with C64x+ cores and offer additional floating-point capabilities for easier high-range computation. There is a "multi-core navigator" which is an evolved DMA system that does not necessitate any intervention from the DSP cores to queue, route and transfer data. This navigator simplifies inter-core communication in addition to communication with coprocessors and external interfaces. A FFT coprocessor accelerates the FFT computations which are numerous in the LTE physical layer computation. The new architecture has an on-chip hierarchical network-on-chip named TeraNet 2, offering a data rate of more than 2 TB per second between cores, coprocessors, peripherals and memory. This network supports several configurations of the operators (cores or coprocessors). The new device contains a shared memory associated with an optimized multi-core memory controller, which is also shared.

The tci6488 and tci6486 can be analyzed using the categories defined in Sect. 4.2. The tci6488 has no internal shared memory; at coarse grain, it can be considered as an MIMD architecture of type NORMA. However, when connected to an external DDR2 memory, it becomes an UMA architecture for this memory because all cores have equal access rights. The tci6486 with its internal shared memory is an MIMD UMA architecture. However, only a small portion of its internal memory is shared; data locality is thus important, even in this UMA architecture. At fine grain, the c64x+ micro-architecture of the cores in both processors is also an MIMD architecture of type UMA. The categories of MIMD and UMA are not sufficient to describe the system-level behavior of a processor for a rapid prototyping tool. They can only focus on some specificities of the architecture without including its global system-level behavior. A model based on operators and their interconnections is necessary to study the possible parallelism of the architecture.

These parallel architectures necessitate rapid prototyping to simulate their behavior and may benefit from automatic multi-core code generation.

5.1.2 Building a New Architecture Model

Among the objectives of this work is to create a graph model for hardware architectures that represent the hardware behavior at a system-level, ignoring the details of implementation while specifying the primary properties. Such a model must be sufficiently abstract to model new architectures including additional features without the need for specific nodes. As the target process is rapid prototyping, the model must be simple to use and understand and must also facilitate multi-core scheduling. Grandpierre and Sorel define in [5] an architecture model for rapid system prototyping using the AAM methodology. Their model specifies four types of vertices: operator, communicator, memory and bus. An operator represents a processing element that can be either an IP or a processor core. A communicator represents a Direct Memory Access (DMA) unit that can drive data transfers without blocking an operator. The model is comprehensively defined in Grandpierre's PhD thesis ([6]). A bus is a communication node for which data transfers compete. As the original model is unable to model switches, Mu extends it in [7], adding IPs and communication nodes that model switches. Switches are bus interconnections that (theoretically) offer perfect contention-free connections between the busses they connect. This is a necessary addition to the model for this study, as the SCR of target processors is a switch. In Sect. 5.2.3, the architecture model of [7] is shown not to be sufficiently expressive to represent the targeted architectures. It also leads to an inefficient scheduling process.

The following sections define a new architecture model named System-Level Architecture Model or S-LAM. This model is an extension of the original architecture model of Grandpierre [5].

5.2 The System-Level Architecture Model

The simplification of the architecture description is a key argument to justify the use of a system-level exploration tool like PREESM over the manual approach. Maintaining high expressiveness is also important because the majority of embedded architectures are now heterogeneous. These observations led to the System-Level Architecture Model described in the following sections.

5.2.1 The S-LAM Operators

An S-LAM description is a topology graph which defines data exchanges between the cores of a heterogeneous architecture. Instead of "core", the term "operator", defined by Grandpierre in his PhD thesis [6], is used. This is due to the fact there is no difference between a core and a coprocessor or an Intellectual Property block (IP) at system-level. They all take input data, process it and return output data after a given time. In S-LAM, all the processing elements are named operators and only their name and type are provided.

Even with several clock domains, execution times must be expressed using one time unit for the entire system. Due to the high complexity of modern cores, it is no longer possible for the clock rate to express the time needed to execute a given actor. Consequently, times in S-LAM are defined for each pair $(actor, operatortype)$ in the scenario instead of using the clock rate of each core. IPs, coprocessors and dedicated cores all have the ability to execute a given set of actors specified in the scenario as constraints. An operator comprises a local memory to store its processed data. Transferring data from one operator to another operator via an interconnection effectively means transferring the data from local memory of one operator to the other. The following sections explain how operators may be interconnected in S-LAM.

5.2.2 Connecting Operators in S-LAM

Two operators may not be connected by merely using an edge. Nodes must be also inserted in order to provide a data rate to the interconnections. The goal of system simulation is to identify and locate bottlenecks, i.e. the particular features that limit the system speed, significantly increase the power consumption, augment the financial costs, and so on. S-LAM was developed to study the particular constraint of speed; reducing this constraint is essential for the study of LTE. Additional features may be needed to explore power or memory consumption. The vertex and edge types in the System-Level Architecture Model, shown in Fig. 5.4, are thus:

- the `parallel node` models a switch with a finite data rate but a perfect capacity to transfer data in parallel. It can be employed to model a switch. As long as a bus does not seem to be a potential bottleneck, it may also be modeled with a parallel node. Parallel nodes reduce simulation complexity as it eliminates the costly transfer ordering step on this communication node,
- the `contention node` models a bus with finite data rate and contention awareness. This node should be used to test architecture bottlenecks,
- the `RAM` models a Random Access Memory, and must be connected to a communication node. Operators read and write data through this connection. An operator has access to a RAM if a set-up link exists between them.
- the `DMA` models a Direct Memory Access. A core can delegate the control of communications to a DMA via a set-up link. DMA must also be connected to a communication node. Delegating a transfer to a DMA will allow the core to process in parallel with the transfer, after a set-up time,
- the `directed (resp. undirected) link` shows data to be transferred between two components in one (respectively both) direction(s),

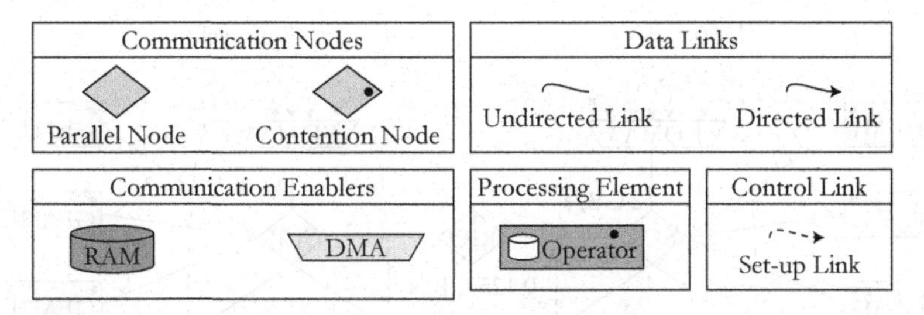

Fig. 5.4 The elements of S-LAM

- the `set-up link` only exists between an operator and a DMA or a RAM. It allows an operator to access a DMA or RAM resource. The set-up link to a DMA provides the time needed for setting-up a DMA transfer.

Both parallel and contention nodes are called Communication Nodes (CN). Directed and undirected links are both called data links. A black dot in a component in Fig. 5.4 shows where contention is taken into account. Contention is expressed as sequential behavior. An operator has contention; it cannot execute more than one actor at a time. A contention node has contention; it cannot transfer more than one data packet at a time. Simulating elements with contention during deployment phase is significantly time-consuming because an ordering of their actors or transfers must be computed. A Gantt chart of the deployment simulation of elements with contention will contain one line per operator and one line par contention node. Connecting an operator to a given DMA via a set-up link requires that this operator allows the DMA the control of its transfers through communication nodes connected to the same DMA via a data link.

5.2.3 Examples of S-LAM Descriptions

An S-LAM description models an architecture behavior rather than a structure. The most logical way to model a tci6488 is to use the EDMA3 to copy data from local memory of one core to the local memory of another. Figure 5.5 shows the S-LAM of two tci6488 with clock rate of 1 GHz connected via their RapidIO links. Since the SCR is a switch element, contention may be ignored on this component because a switch limits much contention. The RapidIO link at 1 Gbit/s $=$ 0.125GByte/s $=$ 0.125 Byte/cycle is likely to be the bottleneck of the architecture. It is represented by a single contention node. Each core operator contains both a C64x+ core and a local L2 memory. These operators delegate both their intra-processor and inter-processor transfers to the EDMA3. The local transfers were benchmarked in [8]. The results were a data rate of 2 GB/s and a set-up time of 2,700 cycles; these are the values used in S-LAM for this study. The Turbo and Viterbi coprocessors, TCP2 and VCP2, each have one local memory for their input and output data.

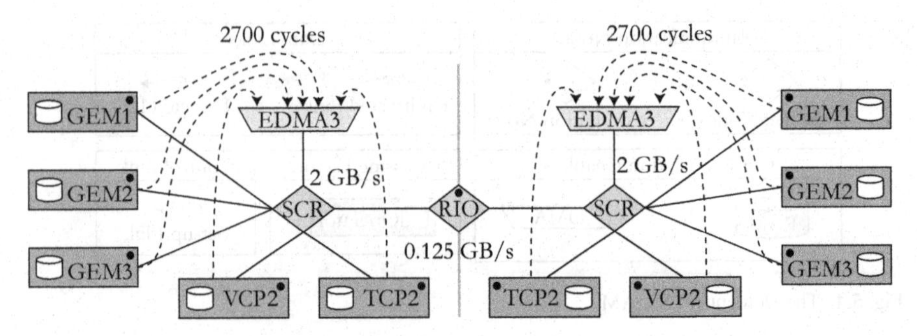

Fig. 5.5 S-LAM description of a board with two tci6488 using EDMA3 for communications local to a processor

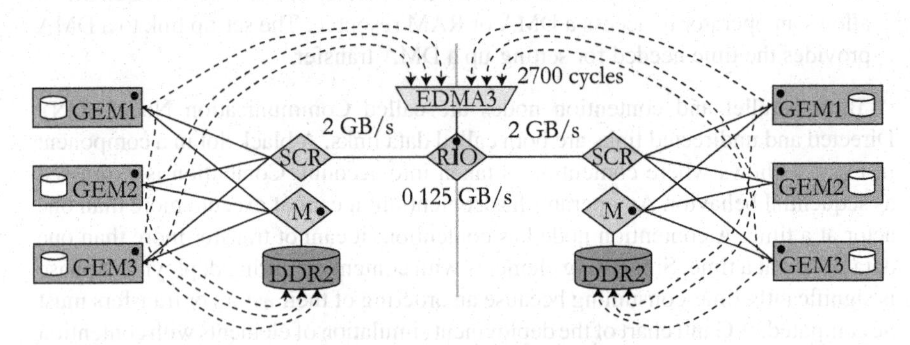

Fig. 5.6 S-LAM description of a board with 2 tci6488 using DDR2 shared memory for communications local to a processor

The simulation of an application on this architecture can be represented by a Gantt chart with 11 lines if all the operators and communication nodes are used. The actors and transfers on the 11 nodes with contention must be ordered while mapping the algorithm on the architecture. This may be contrasted with the architecture model developed by Mu in [7] which computes the contention on each bus and produces a Gantt chart with 24 lines for the same example. The fundamental difference is the ordering process, which is a complex operation and is reserved in the current study for the architecture bottlenecks only. Moreover, the absence of set-up link in Mu's model would make the description of Fig. 5.5 ambiguous where the model identifies the core that delegates its transfers to a given DMA.

The previous model is not the only possibility for a board with two tci6488. The shared DDR2 external memory may also be used when exchanging data between c64x+ cores in a single tci6488 processor. For this case, the S-LAM description can be that of Fig. 5.6. VCP2 and TCP2 have been removed from the block diagram for clarity. The local transfers are much slower when the external memory is used and these transfer easily become an architecture bottleneck. Thus, accesses to the local memory are linked to a specific contention node. Only one EDMA3 node is used to

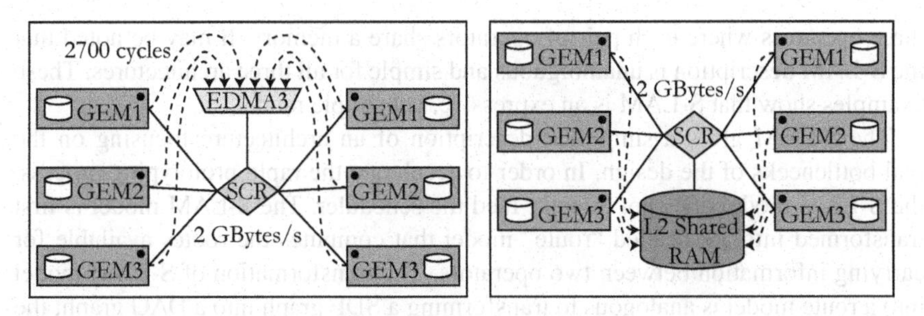

Fig. 5.7 S-LAM description of a tci6486 processor

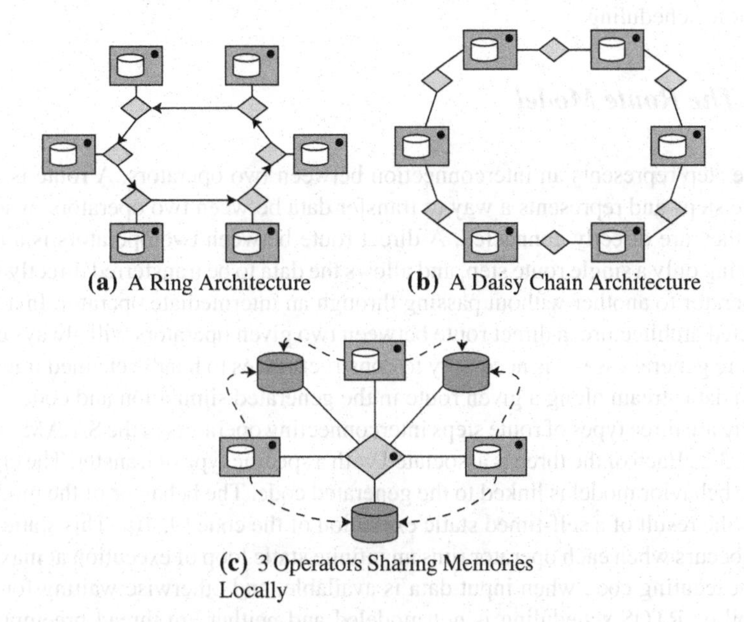

(a) A Ring Architecture **(b)** A Daisy Chain Architecture

(c) 3 Operators Sharing Memories Locally

Fig. 5.8 S-LAM descriptions of architecture examples

model the two DMAs of the two processors. The following example illustrates the abstraction of the S-LAM architecture.

Figure 5.7 shows the two possible models described above for a tci6486 processor. The first uses EDMA3 to copy data from the local memory of one core to the local memory of another. The second uses the shared L2 memory to transfer data. Note that the two media: shared memory and local memory to local memory copy, could be combined; in this case, the SCR node would need to be duplicated to distinguish between the two different route steps, as will be explained in following sections.

As an example of S-LAM expressiveness, Fig. 5.8a shows a unidirectional ring architecture where each operator communicates through a ring of connections. Figure 5.8b shows a daisy chain architecture where each operator can only communicate with its two adjacent neighbors. Figure 5.8c shows an architecture with three operators where each pair of operators share a memory. It may be noted that the S-LAM description is unambiguous and simple for all three architectures. These examples show that S-LAM is an expressive, yet simple model.

The S-LAM gives a simplified description of an architecture focusing on the real bottlenecks of the design. In order to accelerate the rapid prototyping process, the S-LAM model does not directly feed the scheduler. The S-LAM model is first transformed into a so-called "route" model that computes the routes available for carrying information between two operators. The transformation of S-LAM model into a route model is analogous to transforming a SDF graph into a DAG graph; the original model is more natural, expressive and flexible while the second one is better adapted to scheduling.

5.2.4 The Route Model

A route step represents an interconnection between two operators. A route is a list of route steps and represents a way to transfer data between two operators, whether or not they are directly connected. A direct route between two operators is a route containing only a single route step, and allows the data to be transferred directly from one operator to another without passing through an intermediate operator. In a fully connected architecture, a direct route between two given operators will always exist. In a more general case, it is necessary to construct routes to handle chained transfers of the a data stream along a given route in the generated simulation and code.

There are three types of route steps interconnecting operators in the S-LAM model of Sect. 5.2 . Each of the three is associated with a specific type of transfer. The choice of code behavior model is linked to the generated code. The behavior of the modeled code is the result of a self-timed static execution of the code [9, 10]. This static execution occurs when each operator runs an infinite static loop of execution at maximal speed, executing code when input data is available, and otherwise waiting for data availability. RTOS scheduling is not modeled and neither are thread preemptions, nor external event waits. The statically generated code is self-timed and uses data transfer libraries that fit the route step behaviors in Fig. 5.9. Additional modifications may be introduced in the LTE models that make the simulation of external events possible.

The types of route step, as shown in Fig. 5.9 are:

1. The `message passing` route step: the source operator sends data to the target operator via message passing. The data flows serially through one or several route steps before reaching the target. The source and target operators are involved in the process by defining path and controlling the transfer.
2. The `DMA` route step: the source operator delegates the transfer to a DMA that sends data to the target operator via message passing. After a set-up time, the source operator is free to execute another actor during this transfer.
3. The `shared memory` route step: the source operator writes data into a shared memory and the target operator then reads the data. The position of the RAM

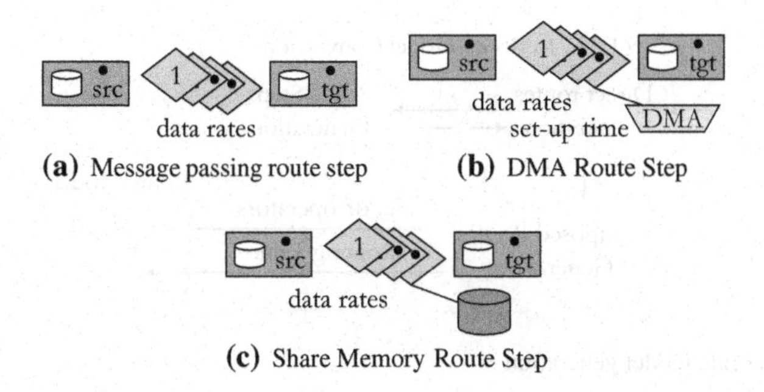

Fig. 5.9 The types of route steps

element (the communication node to which one of the route step communication nodes is connected) is important because it selects the communication nodes used for writing and for reading.

This list of route steps is not exhaustive: there are other data exchange possibilities. For instance, transitioned memory buffers [11] is currently not modeled in this study. In this route step, it is not the data in transitioned memory buffers which is transferred but the "ownership" of this data which is transferred from one operator to another. Thus, it is vital to protect the data against concurrent accesses. Studying such route step would be of interest, as they increase synchronization but can reduce memory needs. Routes can be created from routes steps to interconnect operators. The route model contains two parts: an operator set containing all the operators in the architecture and a routing table, which is a map assigning the best route to each pair of operators $(source, target)$. The route model accelerates the deployment simulations because it immediately provides the best route between two operators without referring to the S-LAM graph. The transformation of S-LAM into route model is explained in next section.

5.3 Transforming the S-LAM Model into the Route Model

S-LAM was developed in order to simplify the multi-core scheduling problem of the PREESM tool. The route pre-calculation that launches this complexity reduction will now be studied.

5.3.1 Overview of the Transformation

The generation of the route model is detailed in Fig. 5.10. Transforming an S-LAM graph into routes is performed in three steps: route steps are generated first, followed by the generation of direct routes and finally by the composed routes. Each step is detailed below.

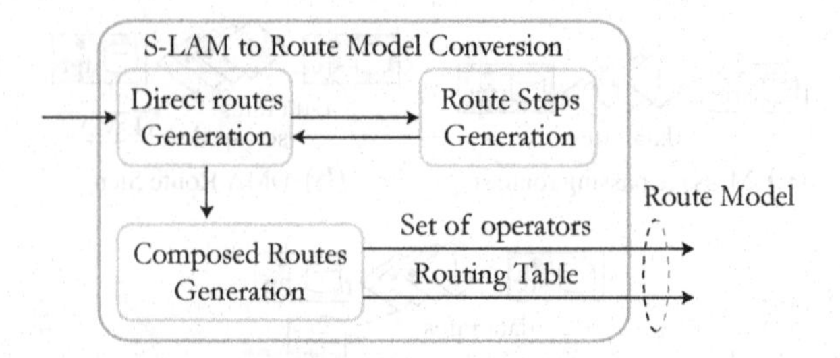

Fig. 5.10 The route model generation

5.3.2 Generating a Route Step

From a source operator, a target operator and a list of communication nodes connecting these two operators, a route step with one of the types defined in Sect. 5.2.4 can be generated. While creating the route step, the communication nodes are scanned and connections to DMA or RAM vertices are searched to determine the current route step type. If a DMA or RAM is found, its incoming set-up links are searched and if these links do not have same source as the current route step, the DMA or RAM is ignored. The advantage of using set-up links is that transfers that are independent of a DMA or a RAM can share a communication node with other DMA-driven and RAM transfers. Contentions between all transfers on a given communication node can be simulated.

5.3.3 Generating Direct Routes from the Graph Model

Using the route step generation function, the direct route generation code parses the graph starting with the source operators src. The algorithm displayed in Algorithms 5.1 and 5.2 scans the communication nodes and maintains lists of previously visited nodes. When a target operator tgt is met, the exploreRoute function in Algorithm 5.2 generates a route step using the method in Sect. 5.3.2. If the new route step has a lower cost than the one (if any) in the table, a new route only containing the new step is stored in a table named routing table. The cost of a route is defined as the sum of the costs of its route steps. The cost of a route step depends on the route step type and is calculated using a typical data size, set in the scenario.

The complexity of the algorithm is $O(PC^2)$ where P is the number of operators in the graph and C the number of communication nodes. This high complexity is not a problem provided architectures remain relatively small. After direct route generation, the routing table contains all the direct routes between interconnected operators. In S-LAM, non totally-connected architectures are authorized. The routes between non-connected operators are still missing at the end of the direct routes generation; they are added during the composed routes generation.

Algorithm 5.1: Direct routes generation

```
  Input: An S-LAM model
  Output: The Corresponding Route Model
1 foreach operator src in operators do
2     foreach interconnection i in outgoing or undirected edges of src do
3         if the edge other end is a communication node n then
4             Add the node n to a list l of already visited nodes;
5             Call exploreRoute(src,n,l);
6         end
7     end
8 end
```

Algorithm 5.2: exploreRoute

```
   Input: An operator src, a communication node n, a list of nodes l
   Output: The best routes from src to its direct neighbors
   /*This recursive function scans the communication nodes and adds a route when reaching a
     target operator                                                                    */
 1 foreach interconnection i in outgoing and undirected edges of n do
 2     if the other end of the edge is a communication node n2 then
 3         Create a new list l2 containing all the elements of l;
 4         Add n2 to l2;
 5         Call exploreRoute(src,n2,l2);
 6     else
 7         if the other end of the edge is an operator tgt then
 8             Generate a route step from src, tgt and the list of nodes l;
 9             Get the routing table current best route between src and tgt;
10             if the new route step has a lower cost than the table route then
11                 Set it as the table best route from src to tgt;
12             end
13         end
14     end
15 end
```

5.3.4 Generating the Complete Routing Table

The routes between non-connected operators are made of multiple route steps. Routes with multiple route steps are built using a Floyd-Warshall algorithm [12] provided in Algorithm 5.3. The route between a source src and a target tgt is computed by composing previously existing routes in the routing table and retaining those with the lowest cost.

The Floyd-Warshall algorithm results in the best route between two given operators with a complexity of $O(P^3)$ and is proven to be optimal for such a routing table construction. The complexity of the routing table computation is not problematic for this study because the architectures are always a small number of cores; the routing table construction of a reasonably interconnected architecture with 20 cores was

Algorithm 5.3: Floyd-Warshall algorithm: computing the routing table

Input: An S-LAM model
Output: The Corresponding Route Model
1 **foreach** *operator k in operators* **do**
2 **foreach** *foreach operator src in operators* **do**
3 **foreach** *foreach operator tgt in operators* **do**
4 Get the table best route from *src* to *k*;
5 Get the table best route from *k* to *tgt*;
6 Compose the 2 routes in a new route from *src* to *tgt*;
7 Evaluate the composition;
8 Compare the table best route from *src* to *tgt* with the composition;
9 **if** *the composition has a lower cost* **then**
10 Set it as the table best route from *src* to *tgt* in the table;
11 **end**
12 **end**
13 **end**
14 **end**

benchmarked at less than 1 s. This may be compared to mapping/scheduling activities for the same architecture which require several minutes. The route model simply consists of this table and the set of operators for the S-LAM model input. If the table is incomplete, i.e. if, in the routing table, a best route does not exist for a pair of operators (src, tgt), then the architecture is considered to be not totally connected via routes. The PREESM scheduler does not handle such non-connected architectures. In this case, PREESM will stop and return an error before starting the mapping and scheduling process. The overall complexity of S-LAM routing is $O(P.(P^2 + C^2))$. In the next section, transfers are simulated in PREESM using routes.

5.4 Simulating a Deployment Using the Route Model

Depending on the route step type, certain transfer simulations are inserted into the execution Gantt chart in addition to actor simulations. The simulation of a data transfer is now described for each type of route step.

5.4.1 The Message Passing Route Step Simulation with Contention Nodes

Part 1 of Fig. 5.11 shows the simulation of a single message passing transfer between actor1 mapped on *src* and actor2 mapped on *tgt*. The transfer blocks the two contention nodes in the route step during the transfer time. The data rate of the transfer (in Bytes per cycle) is the lowest data rate of the communication nodes (which is the bottleneck of this particular communication). Source and target operators are actively transferring the data and are thus unavailable until its completion.

5.4.2 The Message Passing Route Step Simulation Without Contention Nodes

If there is no contention node present, such as in Part 2 of Fig. 5.11, there will not be a line in the Gantt chart for the transfer but the actor2 will be delayed until the transfer is complete. This transfer model is equivalent to those used by Kwok in his mapping and scheduling algorithms [13], where parallel nodes take into account the transfer delays but ignore contentions.

5.4.3 The DMA Route Step Simulation

The simulation of a DMA transfer is shown in Part 3 of Fig. 5.11. The set-up overhead for the transfer is mapped onto the source operator; this transfer is then equivalent to that of single message passing except that the operators are not involved in the transfer. In practice, the overhead corresponds to the set-up time of the DMA, i.e. the time to write the transfer description into the DMA registers.

5.4.4 The Shared Memory Route Step Simulation

The simulation of a shared memory transfer is shown in Part 4 of Fig. 5.11. First, the source operator writes the data to the shared memory and then the target operator reads it. The left-hand side communication nodes are occupied during writing and the right-hand side nodes during reading. The writing and reading data rates can be different. They are both limited by the maximum data rates of the memory and of the communication nodes. This transfer model means that shared data buffers are not used "in place" by the actors. Instead, they manipulate local copies of the data. This choice costs memory for the local copies but can reduce the algorithm latency in certain cases.

5.5 Role of S-LAM in the Rapid Prototyping Process

The S-LAM model was developed to be the input of the rapid prototyping process. It is the S-LAM that is represented by the architecture model block in the diagram shown Fig. 1.2. The typical size of architecture in a S-LAM graph in PREESM is between a few cores and a few dozens of cores.

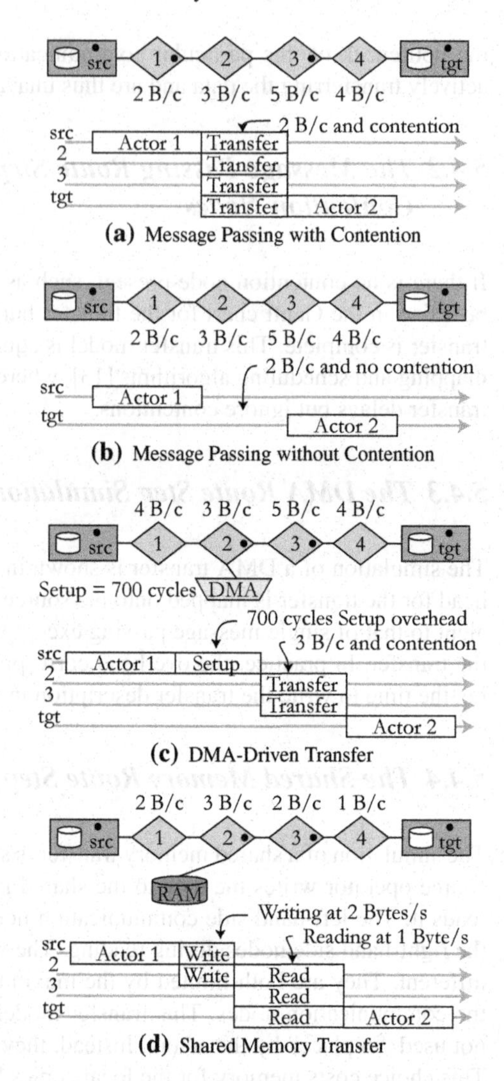

Fig. 5.11 Impact of route types on the simulation of a transfer

5.5.1 Storing an S-LAM Graph

S-LAM consists of components and interconnections, each with a type and specific properties. This model is naturally compatible with the IP-XACT model [14], an IEEE standard from the SPIRIT consortium [14] intended to store XML descriptions for any type of architecture. The IP-XACT language is a coordination language: it specifies components and their interconnection but not their internal behavior. In the IP-XACT organization, a top-level design will contain component instances. These instances reference component files that can contain several views, some ref-

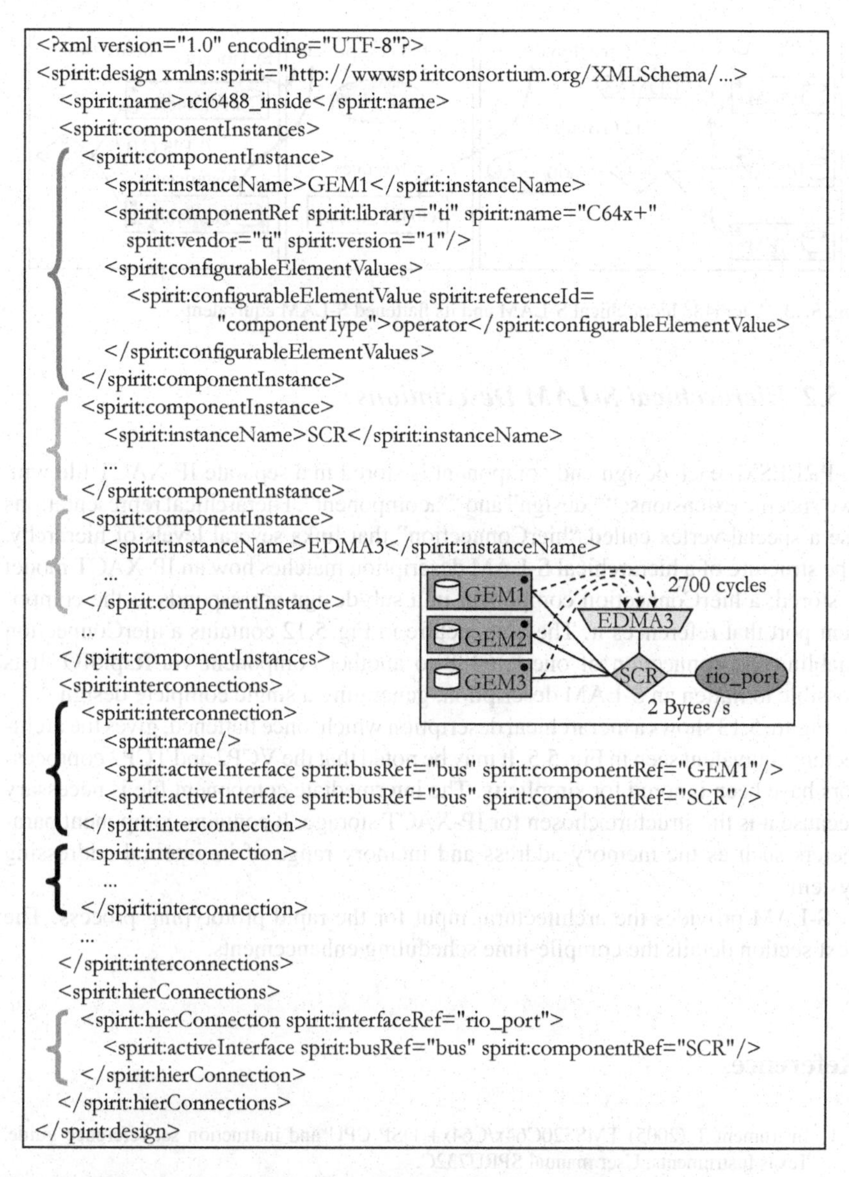

Fig. 5.12 Storing an S-LAM description in an IP-XACT design file

erencing sub-designs. Figure 5.12 displays a simple S-LAM description and part of its corresponding IP-XACT design file.

A very simplified subset of IP-XACT is used where a component can have either one view referencing a single sub-design or no view at all; in that case they are atomic components. It is the existence of sub-designs that makes hierarchical descriptions possible.

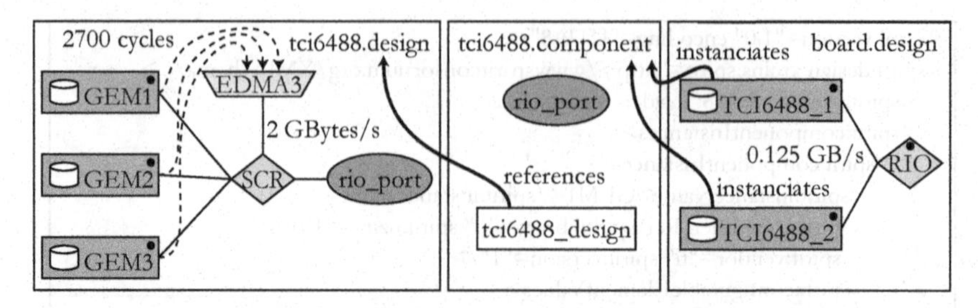

Fig. 5.13 A tci6488 hierarchical S-LAM and its flattened S-LAM equivalent

5.5.2 Hierarchical S-LAM Descriptions

In PREESM, each design and component is stored in a separate IP-XACT file with two specific extensions: "*.design" and "*.component". Hierarchical representations use a special vertex called "hierConnection" that links several levels of hierarchy. The structure of a hierarchical S-LAM description matches how an IP-XACT model is stored: a hierConnection component in a sub-design corresponds to the component port that references it. The architecture in Fig. 5.12 contains a hierConnection enabling the connection of one tci6488 to another component via RapidIO. It is possible to flatten an S-LAM description, generating a single complete design.

Figure 5.13 shows a hierarchical description which, once flattened, gives the architecture equivalent seen in Fig. 5.5. It may be noted that the VCP2 and TCP2 coprocessors have been ignored for simplicity. The intermediate component file is necessary because it is the structure chosen for IP-XACT storage. It contains component parameters such as the memory address and memory range of its internal addressing system.

S-LAM provides the architectural input for the rapid prototyping process. The next section details the compile-time scheduling enhancements.

References

1. Instrument T (2005) TMS320C64x/C64x+ DSP CPU and instruction set reference guide. Texas Instruments, User manual SPRU732C.
2. TMS320TCI6488 DSP platform, texas instrument product bulletin (sprt415) (2007).
3. Sinnen O (2007) Task scheduling for parallel systems (Wiley series on parallel and distributed computing). Wiley-Interscience, Hoboken
4. Friedmann A (2010) Enabling LTE development with TI new multicore SoC architecture SPRY134. Technical Report, Texas Instruments
5. Grandpierre T, Sorel Y (2003) From algorithm and architecture specifications to automatic generation of distributed real-time executives: a seamless flow of graphs transformations. In: MEMOCODE '03, pp 123–132.
6. Grandpierre T (2000) Modélisation d'architectures paralléles hétérogénes pour la génération automatique d'exécutifs distribues temps reel optimisés, Grandpierre, Thierry. http://www.inria.fr/rrrt/tu-0666.html

7. Mu P (2009) Rapid prototyping methodology for parallel embedded systems. Ph.D. thesis, INSA Rennes.
8. Pelcat M, Aridhi S, Nezan, JF (2008) Optimization of automatically generated multi-core code for the LTE RACH-PD algorithm. 0811.0582 (2008), DASIP 2008, Bruxelles, Belgique. http://arxiv.org/abs/0811.0582
9. Sriram S, Bhattacharyya SS (2009) Embedded multiprocessors: scheduling and synchronization, 2nd edn. CRC press, Boca Raton
10. Lee EA (1989) Scheduling strategies for multiprocessor real-time DSP. In: IEEE global telecommunications conference and exhibition. Communications technology for the 1990s and beyond.
11. Bell D, Wood G (2009) Multicore programming guide. Techical Report, Texas Instruments
12. Cormen TH, Leiserson CE, Rivest RL, Stein C (2001) Introduction to algorithms, 2nd edn. The MIT Press, Cambridge
13. Kwok Y (1997) High-performance algorithms of compile-time scheduling of parallel processors. Ph.D. thesis, Hong Kong University of Science and Technology.
14. SPIRIT Schema Working Group (2008) IP-XACT v1.4: A specification for XML meta-data and tool interfaces. Techical Report, The SPIRIT Consortium.

Chapter 6
Enhanced Rapid Prototyping

6.1 Introduction

In Chap. 4, an overview of the multi-core scheduling problem and solutions presented in the literature were summarized. A flexible rapid prototyping process has an important role to play in all the design steps of a multi-core DSP system. This chapter details several methods developed for enhancing, in speed, accuracy and flexibility the rapid prototyping of distributed systems. These methods are used for different phases of the rapid prototyping process, and include a complete separation between algorithm and architecture in description phase (Sect. 6.2), a scheduler structure separating the different necessary heuristics (Sects. 6.3, 6.4 and 6.5) and a clear display of the resulting schedule performances (Sect. 6.6).

6.1.1 The Multi-Core DSP Programming Constraints

Gatherer and Biscondi [1] state that the most important metrics for a high-performance embedded multi-core DSP are, in order of importance, `power, real-time performance, and financial cost`. From a hardware point of view, optimizing a system consists of providing more Million Instructions per Second (MIPS) or Million Multiply Accumulates per Second (MMACs) per Watt to the programmer while maintaining the device cost. From a software point of view, it consists of providing the highest number of possible functionalities for a given hardware while maintaining or reducing software programming costs. Hardware and software codesign consists of jointly optimizing the hardware and the software. The rapid prototyping of an application, such as base station baseband processing, on several architectures performs the important task of optimizing the system parameters. This is sometimes called design space exploration, and is most effective when approached as a codesign, as the optimization of the above three metrics is simultaneous.

M. Pelcat et al., *Physical Layer Multi-Core Prototyping*,
Lecture Notes in Electrical Engineering 171, DOI: 10.1007/978-1-4471-4210-2_6,

The first constraint, `power consumption`, is highly dependent on the architecture and frequency of the target processor. Estimating the limits and advantages of target architectures in early stages of implementation is vital in making the most power efficient choices. Another parameter that influences the power consumption of a multi-DSP system is the locality of data and computation. A globally shared memory is a solution that can not scale to more than a few cores and a distributed memory necessitates an early study of data and code location. Finally, load balancing the DSP cores assists in reducing the chip "hot spots", easing power dissipation and increasing processor reliability. For battery-powered handheld devices, a limit of 1 Watt for the entire system is usual. [2] gives an idea of power dissipation constraints in highly reliable (or carrier-grade) communication systems such as LTE base stations: in an Advanced Telecommunications Computing Architecture (ATCA) rack of carrier cards, each carrier card can dissipate up to 200 Watts and, to obtain a long-term reliable system, each chip on the rack should maintain power consumption under approximately 10 W. This constraint immediately disqualifies general purpose processors which consume tens of Watts.

The second constraint, `real-time performance`, can be defined by two metrics:

- The `execution time` T_e is the minimum period that must separate two consecutive inputs for the system to correctly process all input data. The execution time is the inverse of the `throughput`, which is the maximal input data frequency a system can process.
- The `latency` T_r, also called response time, is the time separating the end of input arrival and the end of its processing. On a multi-core architecture, we have $T_e \leq T_r$.

The reason that the latency is greater than the execution time is that several processed input streams are pipelined, reducing T_e but often increasing T_r. The parallelism exploited by pipelining is called `functional parallelism` [3]. The parameter of execution time is used to evaluate whether an architecture can support the data throughput needed, and the latency must be minimized to ensure the system reacts "soon enough" to a command. For instance, the two-way global delay of LTE is required to stay under 10 ms for a user to have a feeling of good system reactivity. This requirement limits the system latency and parallelism must be exploited to reduce T_r. The portion of algorithm parallelism that can be used to reduce latency is called `data parallelism` [3]. Data and functional parallelism are often combined and intricated in real applications. In general, both data and functional parallelisms must be exploited to obtain an efficient system.

The system `financial cost` depends on many parameters but one important factor in cost reduction is the ability to detect a wrong choice in the development process as rapidly as possible.This detection can be achieved by a high level of programmability and early system simulations. These are both available in the Rapid prototyping process, meaning that this process may have a significant role in reducing the financial cost of a design.

6.1.2 *Objectives of a Multi-Core Scheduler*

The scheduler is the most important part of a multi-core rapid prototyping framework: it is here that the algorithm and architecture are combined. The role of a scheduler is to return an "efficient" multi-core schedule in an "acceptable" time. Schedule efficiency is usually expressed as a `speedup` which is the factor of acceleration from using a multi-core instead of a single core. This notion is introduced for homogeneous architectures and can extend to heterogeneous ones if a main type of core, corresponding to a standard behavior, is chosen. The speedup is then computed only for the cores in the architecture that are of main type. An efficient schedule offers a high acceleration reflected by a high speedup given for the algorithm and architecture constraints. Section 6.6 provides a visual method of assessing a schedule quality in terms of speedup.

In a practical situation of multi-core scheduling, heuristic complexity expressed in asymptotic notation (Sect. 4.4.3) is not sufficient to evaluate the real cost of a scheduling heuristic and so scheduling times which are considered as "acceptable" must be additionally defined. For run-time scheduling, the scheduling deadline is usually less than one second and depends on the real-time constraints of the application. Run-time scheduling for LTE will be discussed in Sect. 8.3. For compile-time rapid prototyping, a scheduling time of a few minutes is acceptable to obtain a rough idea of the system behavior. This means that a programmer can run a schedule and vary its parameters several times in an hour. However, a scheduling time of a few days is acceptable for a schedule of advanced quality compared to the rapid prototyping schedule.

A high flexibility is essential for the rapid prototyping process. The following sections define features which serve to enhance this process.

6.2 A Flexible Rapid Prototyping Process

Each implementation has specific constraints and objectives. A rapid prototyping method must adapt to these needs; this is the goal of the flexible rapid prototyping process presented in this document. In Fig. 1.2 of the introduction, a single matching node was seen to be the convergence point of the algorithm and the architecture. The internal functionalities of this matching node in the presented framework are displayed in Fig. 6.1. Prior to multi-core scheduling, the algorithm and the architecture are transformed to extract their parallelism adequately for the scheduling. The input algorithm is described in an IBSDF graph and the architecture in an S-LAM graph. An additional input named scenario is generated, and controls the prototyping parameters. Additional to simulation and code generation, internal data, including algorithm, architecture or schedules, can be exported in files that feed other tools. All the functionalities in the process are optional and may be combined in graphical

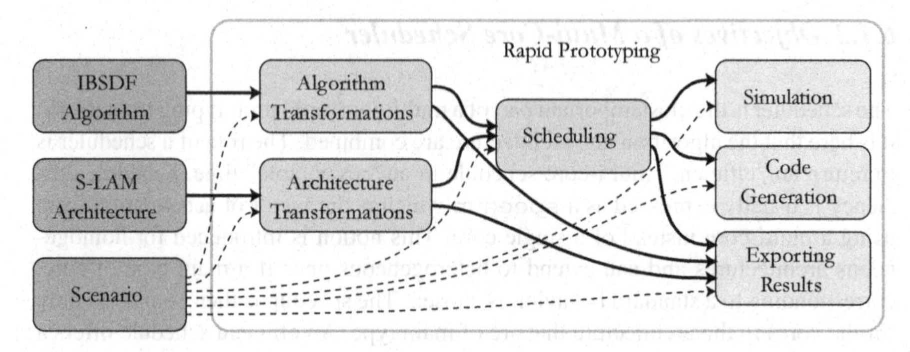

Fig. 6.1 Overview of the rapid prototyping process

workflows to provide advanced flexibility to the programmer. These elements are explained in the following sections.

6.2.1 Algorithm Transformations While Rapid Prototyping

Due to its advantages of hierarchy, predictability and parallelism expressiveness, IBSDF is the chosen algorithm input of the rapid prototyping process and was proposed by Piat in [4]. The top level of its hierarchy contains the actors to schedule. The top level is converted into a DAG before scheduling (Sect. 3.2.3). The degree of graph flattening influences the resulting schedules as does the time necessary to obtain them. Some special vertices which are introduced in the model are now described.

The typical number of actors for scheduling in PREESM is between one hundred and several thousand. The IBSDF model was developed to describe an application consistent with this size for the rapid prototyping process. With IBSDF, sub-graphs can be combined to build a hierarchical application. The strength of this system is that the behavior of a sub-graph can not make the top graph behave unexpectedly, and the schedulability can be checked at compile-time, independently for each hierachy level. The model is constructed so that the `interface` of an actor, i.e. the number of its input and output edges and their sizes, remains unmodified when the graph is transformed. For instance, converting the SDF graph of Fig. 6.2 to a single rate SDF graph (Sect. 3.2.2) adds fork and join vertices that are only used to split or gather data between edges. The edges going from and to A, B, C and D actors are unchanged between the SDF and single rate SDF graphs. Protecting the interface of an actor is important because it is necessary to link this interface to a host code and this host code needs to know where to retrieve or send the data tokens and whether the data is being sent or received.

PREESM is the rapid prototyping tool that serves as a laboratory for the methods presented in this document. The actor interfaces are specified in PREESM using the

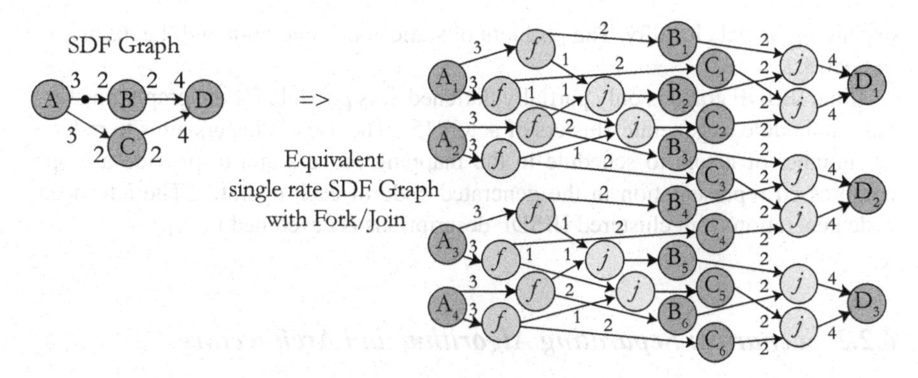

Fig. 6.2 Creating a single rate SDF with fork and join actors

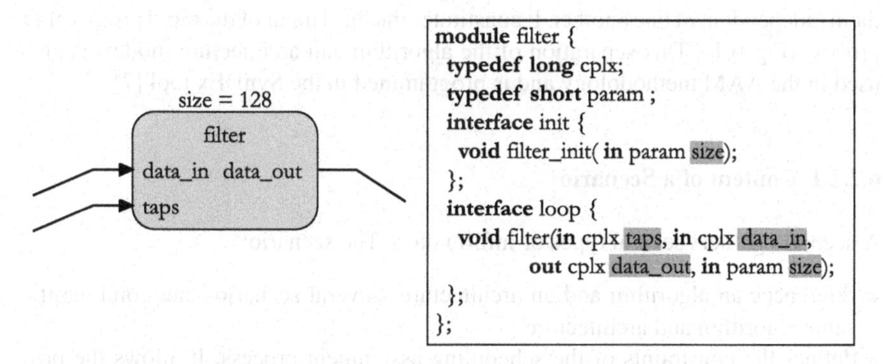

Fig. 6.3 Example of an IBSDF actor with ports and its IDL prototype

generic CORBA Interface Description Language (IDL). Each actor is associated with a loop function prototype defined in an IDL file that specifies the right parameters in the right order. Ports are added to the IBSDF graph to identify the edges and port names are referenced in the IDL files. In Fig. 6.3, the prototype of a filter function is displayed. The data productions and consumptions (in tokens) are specified in the graph and not in the IDL file. The input "size" here is a constant value set in graph parameters; IDL can reference constant values additionally to edges. As IDL is language-independent, the coordination code as well as its stored parameters and actor interfaces are then independent of the host code.

Additionally to the loop function prototype, the IDL file provides the generic prototype of the initialization function call associated with an actor. Initialization functions are used to create initial tokens in the graph whereas loop functions are called at each iteration of the actor.

Sometimes it is necessary to broadcast the same data to several receiving actors in a configurable way. The broadcast actor was introduced in Sect. 3.5.2 to solve hierarchical flattening problems. When generating imperative code from dataflow models, broadcasts can be implemented by memory copies or pointer references depending on data locality. The problem of static code generation will be discussed in Sect. 8.2.

If the IBSDF graph is only partially flattened, it is possible for the graph transformation module to generate clusters of actors [5]. The use of clusters aims to reduce the number of actors to schedule in the mapping process and to provide a loop-compressed representation in the generated code of each operator. The advanced code generation from clustered IBSDF descriptions is developed in [6].

6.2.2 Scenarios: Separating Algorithm and Architecture

The role of a scenario is to separate the algorithm and architecture models, making them independent of one another. It constitutes the third input of the rapid prototyping process (Fig. 6.1). This separation of the algorithm and architecture models is also used in the AAM methodology and is programmed in the SynDEx tool [7].

6.2.2.1 Content of a Scenario

A scenario gathers several types of information. The scenario:

- references an algorithm and an architecture. Several scenarios can combine the same algorithm and architecture.
- defines the constraints of the scheduling assignment process. It allows the programmer to fix the assignment of certain actors to given operators. This way, a hardware coprocessor may be defined as dedicated to certain actors and can not execute other actors. As IBSDF algorithms are hierarchical, these assignments are hierarchical, and apply to all actors including those that contain a graph with other actors.
- associates reference times with each couple (*actor*, *operator type*).
- parameterizes the implementation simulation.
- assigns values to algorithmic graph parameters. The programmer can thus switch between several configurations and prototype several test cases simultaneously.

The advantage of switching scenarios is that a programmer can easily test one algorithm on several architectures, several algorithms on one architecture or explore several sets of parameters.

6.2.2.2 Application Timing on a Target Architectures

In Dataflow MoCs, the actors are not timed. The only relation considered between actors is causality, i.e. who precedes whom. However, the primary constraint of the LTE application is latency. Each couple (*actor*, *operator type*) must be associated with a time representing its behavior. This time is saved in the scenario to protect the algorithm abstraction.

The execution time of a software actor is usually complex and variable. The targeted algorithm granularity describes an entire application with approximately one thousand actors. This implies that the actors usually contain conditioning (i.e. there are “if” or “branch” constructs in the host code). As there are several execution possibilities, an actor with conditioning can not be deterministically timed at compile-time. Moreover, dataflow descriptions do not include the environment in the model of data or control exchanges. Actors are usually included in dataflow descriptions that wait for an external event before executing code or sending token. These actors naturally have unpredictable execution times. However, some execution test cases can be timed using a profiler or instrumenting the software to collect timestamps.

More unpredictability is generated from the hardware. Most general purpose processor cores can execute Out-of-Order (OoO) instructions, changing the instruction execution order at run-time depending on input data availability. A DSP core like the c64x+ (Sect. 5.1.1) has no such capacity, making it more predictable than general purpose processors. However, it has features that complicate predictions:

- The 11-stage pipeline of the c64x+ core greatly complicates the cycle-accurate time prediction of the different cases of a code with conditioning.
- The L1 cache with automatic coherence management is another source of prediction complexity. Certain actor execution orders can cache the data at more appropriate times than other actor execution orders. The actor time can be evaluated under favorable cache conditions (warm cache) and unfavorable cache conditions (cold cache).
- When using external DDR2 memory, a part of the internal L2 memory can serve as a cache for external accesses, again adding unpredictibility.

Despite these variations, a programmer can evaluate system-level behavior of an implementation from a typical execution time, known as Deterministic Actor Execution Time (DAET). DAET can either be Worst-Case Execution Times (WCET) for testing fixed deadlines or warm and cold cache average times to test the typical system behavior in favorable and unfavorable conditions. The type of DAET is influenced by the real-time category of the system. The result of a hard real-time computation is considered useless if it is returned too late while the result of a soft real-time system provides decreasing service quality when its lateness increases. LTE is a hard real-time system. Its final implementation must be tested under WCET conditions. However, average times are used in this study which focuses on the early design process and where it is the system-level behavior which is of interest.

The time unit is not specified in the scenario. Time is a natural integer and it is the role of the programmer to choose a time quantum. Evidently, an easily manipulatable time quantum is desirable; 1 ns is usually employed as it has a relationship with the clock period. For example, this quantum is natural for the tci6488 at 1 GHz because it corresponds to the CPU clock period. To prototype the tci6486 at 500 MHz, a time quantum of 2 ns, the CPU clock period or 1 ns, which is half the clock period, are obvious choices.

6.2.2.3 The Scenario Simulation Parameters

Several simulation parameters are set in the scenario. They include:

- `Token types`: in the application graph, edges carry tokens, which are an abstract data quantum. Each edge has a token type, for example, "cplx" for complex value. This token type must remain abstract for the application to be combined with an architecture. Thus, the scenario protects the algorithm MoC abstraction, defining a token type size for each data token. Like the time quantum, the token size unit is abstract. For example, the programmer can choose a unit size of 1 Byte or 1 kBytes. The usual unit for the target architectures of this study is 1 Byte. The complex symbol values in LTE are usually stored as one 16-bit real part and one 16-bit imaginary part. In this case, the size of *cplx* is then 4.
- `Main operator and Communication Node` (CN): the scenario associates a main operator and a main CN (Sect. 5.2) to the architecture. These are the elements that are primarily studied during scheduling. The schedule quality assessment chart, as explained in Sect. 6.6, evaluates the execution speedup of operators in the S-LAM architecture of the same type as the main operator. It may be noted that the main operator and communication node are the components which are subject to the greatest number of optimizations by the scheduling algorithms.

In PREESM, time and constraint information can be imported in a scenario from Excel sheets. A programmer can generate these times from formulas or benchmarks and import them automatically. It is the existence of such implementation details that increases the ease and rapidity of use of a development chain.

6.2.3 *Workflows: Flows of Model Transformations*

Workflows are graphically edited graphs which represent the successive transformations necessary for the input models to simulate or generate executable code. Workflows are used in the PREESM tool to tune the rapid prototyping process. The three kinds of graphs, algorithm, architecture and workflow, are edited though use of the same generic graph editor named Graphiti [8]. Unlike when working with scenarios a programmer can use the same workflow to prototype several applications and architectures. Two common workflows will be presented below.

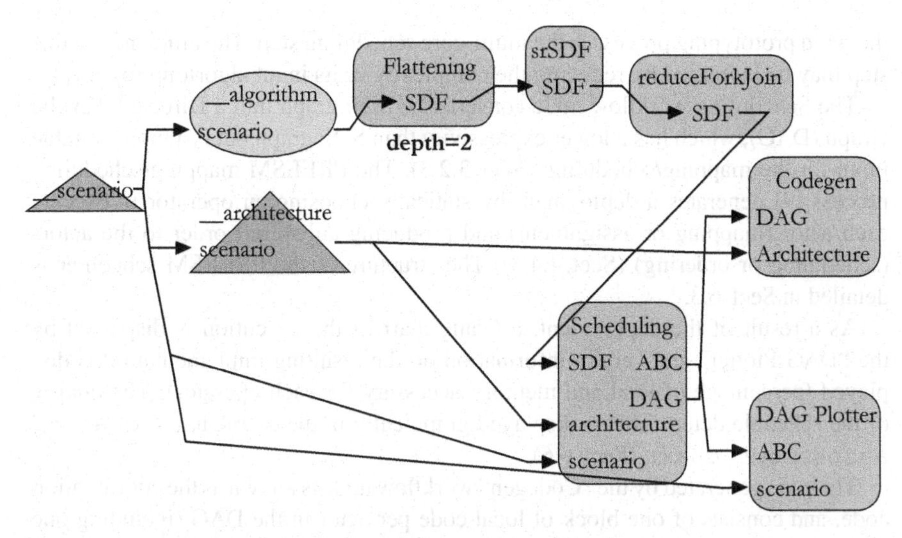

Fig. 6.4 A workflow for prototyping an application

6.2.3.1 A Workflow to Prototype an Application

Rapid prototyping consists of simulating an implementation and then extracting information from this simulation. The workflow in Fig. 6.4 prototypes the three elements (*scenario*, *algorithm*, *architecture*).

A scenario outputs the algorithm and the architecture, and is the only node without an input edge. The shape of the workflow graph is intuitive, as a scenario always references one specific algorithm and one specific architecture. A workflow is applied to a scenario as this scenario initially contains or references all information necessary for the rapid prototyping process. The output of the algorithm node is an IBSDF graph and the architecture node outputs an S-LAM graph.

In the Fig. 6.4, three transformations are applied to the IBSDF graph. The two highest levels of hierarchy are first flattened by the "Flattening" node with parameter $depth = 2$. Its top level is then converted into a single rate SDF graph (Sect. 3.2.2) by the single rate SDF transformation node. The single rate SDF transformation provides the scheduler with a graph of high potential parallelism as all vertices of the SDF graph are repeated according to the SDF graph's Basic Repetition Vector. Consequently, the number of vertices to schedule is greater than in the original graph. Additionally, the single rate SDF conversion is likely to introduce multiple Fork and Join actors (Sect. 6.2.1). The third transformation is applied by node "reduceForkJoin" and minimizes the number of these spurious actors.

The purpose of these transformations is to reveal the potential parallelism of the algorithm and to simplify the work of the actor scheduler. The programmer can tune the depth parameter of the hierarchy flattening node, choosing between a highly parallel implementation and a fast scheduling process. The most complex phase of the rapid prototyping process is the multi-core scheduling step. The efficiency of this step may be increased by reducing the complexity of its input algorithm top level.

The Scheduling workflow node converts the SDF graph into a Directed Acyclic Graph (DAG), which has a lower expressivity than SDF graph but is a more suitable input for the mapping/scheduling (Sect. 3.2.3). The PREESM mapping/scheduling process [9] generates a deployment by statically choosing an operator to execute each actor (mapping or assignment) and producing an overall order to the actors (scheduling or ordering) (Sect. 4.4.3). The structure of the PREESM scheduler is detailed in Sect. 6.3.

As a result of the deployment, a Gantt chart of the execution is displayed by the "DAG Plotter" and certain information on the resulting implementation is displayed (percentage of load and memory necessary for each operator). The quality of the schedule determined is displayed graphically in the `schedule quality assessment chart` (Sect. 6.6).

The code generated by the "Codegen" workflow node is known as the coordination code, and consists of one block of local code per actor in the DAG (including one function call for an actor with no hierarchy), a static schedule of the actors for each processor, and data transfers and synchronizations between the processors. The coordination code is first generated in a generic imperative XML format and then converted into a host code (C code for this study) by an Extensible Stylesheet Language Transformation (XSLT [10]). The host code is hand-written. Static code generation will be further explained in Sect. 8.2 when it is applied to the LTE random access algorithm.

6.2.3.2 A Workflow Combining Rapid Prototyping with SystemC Simulations

Accurate deployment simulations can be generated from the deployment through the use of SystemC-based [11] simulator. This simulator, developed by Texas Instruments, is not publicly available. The workflow, shown in Fig. 6.5, exports the DAG of its execution, produced by the scheduling node, in a graphml format. From this DAG, a XSL transformation can generate a XML or a text file. The SystemC simulation process requires two inputs and both are generated from the DAG using two different XSL files. In Fig. 6.5, it may be seen that two specific files are generated that feed the SystemC rapid prototyping process: one LUA file and one GraphML file. Graph transformations, identical to those shown in Fig. 6.4 could be used to prepare the algorithm graph before scheduling (hierarchy flattening...) in this workflow.

The rapid prototyping process and the SystemC simulator are highly complementary. The rapid prototyping process necessitates a simplified model of the architecture behavior. The S-LAM model contains a suitably simplified view to allow the acceleration of the rapid prototyping process. The SystemC simulator has a complete model of the architecture but has no automatic assignment heuristic and, without a rapid prototyping tool, actor assignments must be processed manually. Consequently, it can compute a cycle-accurate simulation of the implementation and check the accuracy of the rough simulation performed during rapid prototyping.

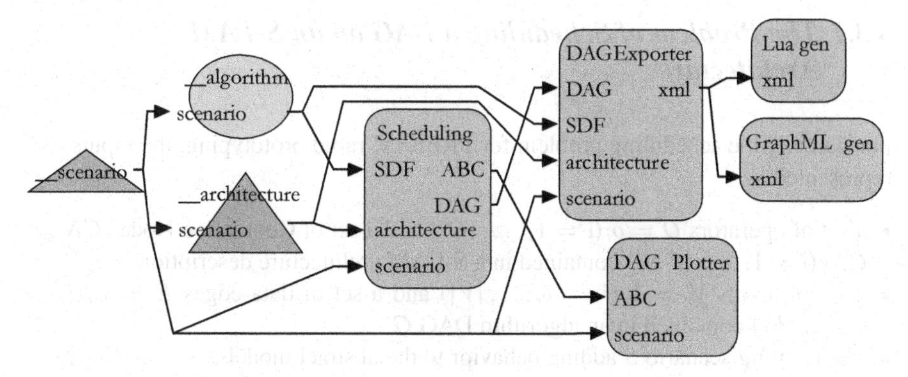

Fig. 6.5 A workflow combining rapid prototyping with SystemC simulations

For example, the EDMA3 module in the tci6488 processor (Sect. 5.1.1) has six Transfer Controllers (TC), and each TC is able to packetize and transfer data in parallel. The EDMA3 is a master of the Switched Central Resource (SCR) which has specific bus configurations for each connected element. The resulting system behavior is highly complex. In Fig. 5.5, the SCR is represented by a parallel node driven by the EDMA3. In LTE applications, the rapid prototyping S-LAM model generates latency estimates with an error of less than 10 % compared to the calculations of the more complete EDMA3 model of the SystemC simulator. The pairing of a fast simulation for prototyping and a precise simulation to evaluate the resulting system favors efficient design choices at reduced costs. The SystemC accurate simulation can be used to evaluate the scheduler performance.

Moreover, only connected IBSDF graphs may be checked for schedulability with the method described in Sect. 3.2. Consequently, several unconnected applications that share a multi-core architecture are not within the scope of current work. However, several unconnected applications can be checked for schedulability and scheduled separately and then combined to be concurrently simulated in a SystemC simulator.

6.3 The Structure of the Scalable Multi-Core Scheduler

As explained in Chap. 4, scheduling is a complex process for which heuristics must be finely tuned to provide good schedules at reduced cost.

6.3.1 The Problem of Scheduling a DAG on an S-LAM Architecture

Formalizing the scheduling problem for PREESM rapid prototyping, the inputs are represented as:

- a set of operators $O = o_i(i = 1, \ldots, |O|)$ and a set of Contention Nodes $CN = CN_i(i = 1, \ldots, |CN|)$ contained in a S-LAM architecture description,
- a set of actors $V = V_i(i = 1, \ldots, |V|)$ and a set of data edges $E = E_i(i = 1, \ldots, |E|)$ contained in an algorithm DAG G.
- a scheduling scenario S adding behavior to the abstract models.

The input to the algorithm is a graph where concurrency between actors is expressed. Consequently, the parallelism extraction phase, illustrated in Fig. 4.2, is omitted from the multi-core scheduling process. Moreover, as the assignment is static, the generated code may be self-timed, avoiding unnecessary scheduling costs; the timing phase is thus transparent because it only depends on actor availability. Two operations remain in the scheduling process:

- `assignment`: each actor is assigned an operator and each edge is assigned a set of CN.
- `ordering`: the actors are processed by one operator and the edges processed by each CN are ordered.

If no CN is defined or if transfer competitions are ignored, the scheduling problem is reduced to actor scheduling. Otherwise, transfers must also be scheduled on routes retrieved from the S-LAM route model (Sect. 5.2.4). In next section, the structure of a scalable scheduler is described. This scheduler separates the different problems of scheduling.

6.3.2 Separating Heuristics from Benchmarks

The scheduler architecture presented in this section is implemented in the PREESM tool; thus it is called the PREESM scheduler. However, underlying method can be applied generally:

- a scheduler intended for rapid prototyping should offer scalable schedule accuracy and scalable scheduling time,
- such scalable behavior may be extended if the assignment heuristic is separated from the architecture benchmark computer.

All scheduling heuristics are based on the same principle: the heuristic takes certain assignment and ordering decisions, the resulting implementation cost is computed, and then this cost is used to determine subsequent assignments. In the literature,

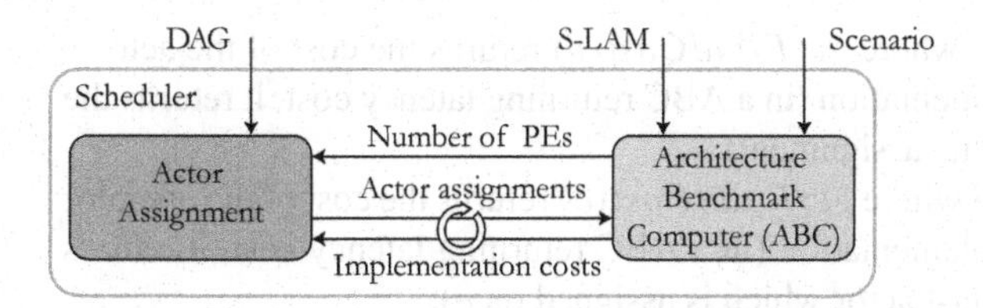

Fig. 6.6 The scheduler sub-modules: actor assignment and ABC

all algorithms embed both assignment decisions and implementation cost evaluation. Moreover, if the minimized cost is the total execution latency (which is the most common case), a cost evaluation requires both the actors on operators and the edges on the CNs to be ordered. The PREESM scheduler splits these functionalities into two sub-modules which share minimal interfaces: the `task assignment` sub-module and the `Architecture Benchmark Computer` (ABC) sub-module. The task assignment sub-module assigns actors to operators and then queries the ABC sub-module, which then evaluates the cost of the proposed solution.

At heuristic initialization, the ABC sub-module transmits the number of operators available to the actor assignment sub-module. Next, the actor assignment sub-module assign actors to operators, and communicates these actor assignments to the ABC sub-module, which then returns the associated cost (infinite if the deployment is impossible). This process is illustrated in Fig. 6.6. One advantage of this approach is that any task assignment heuristic may be combined with any ABC sub-module, leading to many different scheduling possibilities. For instance, an ABC sub-module minimizing deployment memory or energy consumption can be used without modifying the task assignment heuristics. The new sub-module will only return a cost of a different type. The S-LAM architecture is an input of the ABC sub-module (Fig. 6.6). Another advantage is that the assignment heuristic can be architecture-independent. Indeed, it bases its assignment choices only on abstract costs and the number of operators available.

The interface between the ABC sub-module and the actor assignment sub-module consists of:

- $assign : V, O \rightarrow \varnothing$ where $assign(v, o)$ assigns the actor v to the operator o,
- $free : V \rightarrow \varnothing$ where $free(v)$ breaks the assignment of the actor v,
- $getLocalCost : V \rightarrow \mathbb{N}$ where $getCost(v)$ returns the cost of the actor v assignment alone (in an ABC returning latency cost, it returns the execution time of v given its assignment),
- $getLocalCost : E \rightarrow \mathbb{N}$ where $getCost(e)$ returns the cost of the edge e assignment alone (in an ABC returning latency cost, it returns the transfer time of e given its assignment),
- $getCost : I \rightarrow \mathbb{N}$ where $getCost(i)$ returns the cost of the implementation i (in an ABC returning latency cost, it returns the global latency of the implementation). If certain assignments are not chosen, the implementation is incomplete and the ABC will return ∞ if it can not evaluate the cost of a partially assigned implementation.

- $getFinalCost : V \rightarrow \mathbb{N}$ where $getFinalCost(v)$ returns the cost of the actor v assignments in the implementation (in a ABC returning latency cost, it returns the finishing time of v given its assignment),
- $getFinalCost : O \rightarrow \mathbb{N}$ where $getFinalCost(o)$ returns the cost of the operator o assignments in the implementation (in a ABC returning latency cost, it returns the finishing time of the last actor which is assigned on o),

The ABC is free to to return abstract costs of any type, including memory costs, energy costs, execution time. The ABC interface with the assignment heuristic provides the information necessary for both partial implementation and total implementation evaluations. The next section details the actor assignment heuristics and ABC sub-modules that have been of interest to this study.

6.3.3 Proposed ABC Sub-Modules

During a scheduling process, an ABC sub-modules initially receives the S-LAM architecture description and the scenario. It is then responsible for assigning a cost to the schedules it receives. The primary constraint of LTE algorithms is latency. Several ABCs have thus been developed to minimize this parameter, evaluating the implementation latency in different cases of execution and with scalable precision, reusing the concept of time scalability introduced in SystemC Transaction Level Modeling (TLM) [12]. These sub-modules are called latency ABCs. SystemC TLM defines several levels of temporal system simulations, from untimed to cycle-accurate precision. This concept been extended to the development of several ABC latency models with different time precisions. Currently, the types of coded latency ABCs are:

- The `loosely-timed ABC` that accounts for task times of operators and transfer costs on PN and CN. However, it does not consider transfer contention (ignoring the difference between Parallel and Contention Nodes in the S-LAM architecture).
- The `approximately-timed ABC` that associates each inter-core contention node with a constant rate and simulates contentions on CNs.
- The `accurately-timed ABC` that includes the set-up time necessary to initialize a parallel transfer controller such as Texas Instruments Enhanced Direct Memory Access (EDMA [13]). This set-up time is scheduled in the core which triggers the transfer. Accurately-timed latency ABC executes the S-LAM simulation presented in Sect. 5.4.
- The `infinite homogeneous ABC` which is a special ABC that performs an algorithm execution simulation on a homogeneous architecture containing an infinite number of cores with main type. It may be noted that for this study, the main core type of an S-LAM architecture is defined in the input scenario. This ABC enables the extraction of the critical path of the graph (Sect. 4.4.3). The use of infinite homogeneous ABCs is detailed in Sect. 6.4.2.

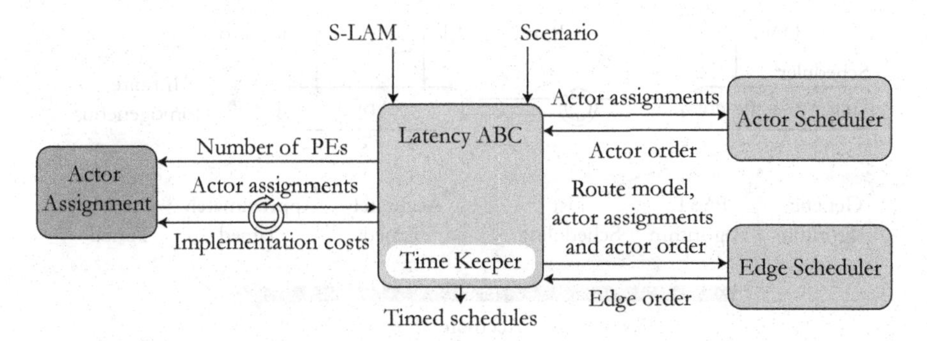

Fig. 6.7 Structure of a latency ABC

All these latency ABCs have the capability to `balance the loads` of the system. Joint latency minimization and load balancing is explained in Sect. 6.4.3.

One role of latency ABCs is to to order actors on operators and (possibly) edges on CNs. Ordering elements while minimizing constraints is equivalent to single-core scheduling, which is a NP-complete problem. Latency ABCs delegate actor and edge ordering to ordering heuristics that can be chosen independently. Up to three separate heuristics may used simultaneously: assignment, actor ordering and transfer ordering. Latency ABCs involve complex mechanisms to insert and remove actors corresponding to transfers. When needed, a time keeper calculates t-level and b-level of each actor. T-level and b-level are needed for list scheduling (4.4.3) and for schedule latency evaluation. The structure of latency ABCs is displayed in Fig. 6.7.

When a data token is transfered from one operator to another, transfer actors are created and then mapped to the CNs of the chosen route. A route may pass through several other operators (Sect. 5.2). If invoked, the edge scheduling sub-module orders the route steps on CNs. Edge scheduling can be executed with algorithms of varying complexity, which results in another level of scalability. The primary advantage of the scheduler structure is the independence of scheduling algorithms from cost type and benchmark complexity. Section 6.4 demonstrates how latency ABCs conform to the scheduler architecture and provide advanced benchmarking.

6.3.4 Proposed Actor Assignment Heuristics

The behavioral commonality of the majority of scheduling algorithms resulted in the choice of the scheduler module structure. Currently, three algorithms are coded in PREESM and are modified versions of the algorithms introduced by Kwok and described in Sect. 4.4.3: a list scheduling algorithm, the FAST algorithm with its parallel version PFAST and a genetic algorithm. Figure 6.8 shows several different assignment heuristics and latency ABCs. It may be noted that having a choice between

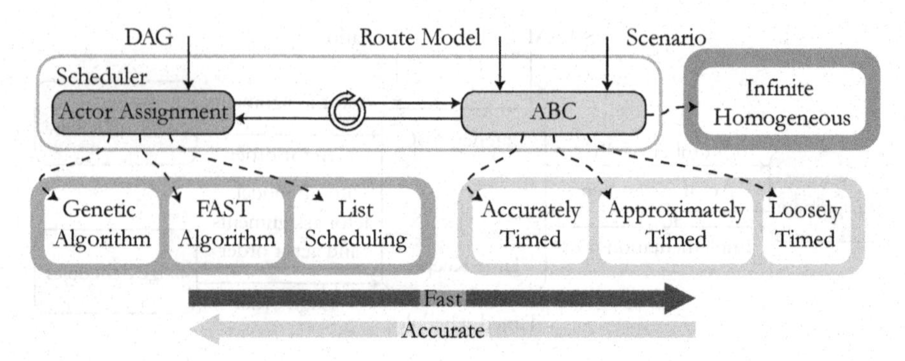

Fig. 6.8 Assignment and existing versions of ABCs

three of each sub-module category, the result is nine possible tradeoffs between precision and time.

Combining heuristics and ABCs, experiments may be performed on the transformed versions of the list, and the FAST and genetic algorithms from Sect. 4.4.3. The original heuristics in the literature were algorithms limited to the Bounded Number of Processors (BNP) while the extended models are used for the architecture heterogeneity of Arbitrary Processor Networks (APN) described in S-LAM. The next section explains how ABCs and heuristics are integrated into the global schedule process.

6.4 Advanced Features in Architecture Benchmark Computers

ABCs link the S-LAM model and the actor assignment process. The next section demonstrates how S-LAM and route models connect to the ABC and then details certain advanced features of latency ABCs.

6.4.1 The Route Model in the AAM Process

Simulating data transfers during mapping and scheduling makes the deployment simulations more complex because vertices representing transfers are dynamically added to and removed from the graph. The addition of new cores to the architecture should not increase the mapping and scheduling complexity exponentially. To achieve this goal, the S-LAM is transformed into the route model before the mapping and scheduling procedures. The resulting scheduling time-complexity on a ring architecture was shown to be linear in [14]. The process of transforming the S-LAM model into a route model was presented in Sect. 5.3.

6.4.2 The Infinite Homogeneous ABC

The infinite homogeneous ABC is unique because its intended use is to extract general information from the algorithm rather than assisting assignment heuristics, which is the general purpose of latency ABCs. This ABC enables Unbounded Number of Clusters (UNC) studies of an implementation; these studies are useful in computing the span and the work of the algorithm under certain constraints, in addition to the critical path, where:

- The `critical path` is the longest path in the whole acyclic graph. To calculate, a typical Communication Node needs to be defined with a given data rate and a typical operator type with DAET for each actor. The main CN and main operator are chosen in the scenario. The critical path can then be computed as presented in Sect. 4.4.3.
- The `span` is the critical path when communication cost is ignored, i.e. in the UNC case. Its length corresponds to an ideal minimal latency on an infinite homogeneous architecture with infinitely fast media.
- The `work` is the execution time of the entire graph on one operator with typical operator type. It is the sum of actor DAETs.

To study these properties, only data dependencies between actors need to be considered and actors need not be ordered on their operators. The infinite homogeneous ABC enables the creation of the CPN dominant list necessary for the list scheduling described in Sect. 4.4.3. The name of this ABC can be misleading: It does not require an infinite architecture, but an architecture with sufficent operators in O to execute as many actors in parallel as possible. There exists O_{min} so that if $|O| \geq O_{min}$, the latency of the schedule is equal to the critical path length and $O_{min} \leq |V|$ because with more than $|V|$ operators, each actor can be assigned to a different operator and adding operators can not reduce the critical path.

6.4.3 Minimizing Latency and Balancing Loads

Load balancing is an important property for a schedule. This feature spreads the processing load and the power dissipation over all operators and reduces the temperature of the system. All preceding techniques have focussed on minimizing the latency. This section presents a method that jointly optimizes latency and load balancing. Both parameters, latency and load balancing, will be shown to be equivalent in the Unbounded Number of Clusters (UNC) case and non-equivalent in Bounded Number of Processors (BNP) and Arbitrary Processor Network (APN) cases (Sect. 4.4.3). A method is then proposed to allow joint optimization in BNP and APN cases. This method maybe activated for any latency ABC, transforming the returned cost.

6.4.3.1 Equivalence of Problems in UNC Case

Minimizing latency and balancing loads are quite 'similar' problems. Indeed, in the little constrained UNC scheduling case (Sect. 4.4.3), the following theorem states:

Theorem 6.4.1 *When scheduling a DAG on an Unbounded Number of Clusters (UNC), the problem of balancing the loads is equivalent to the problem of minimizing latency.*

Proof We have:

- $\mathbf{V}$ a vector containing all the actors $v_i \in V$, with V the actor set of the DAG,
- $\mathbf{O}$ a vector containing all the operators executing at least one actor in the operator set O,
- $P = |\mathbf{O}| \leq |\mathbf{V}|$ the number of operators executing at least one actor,
- $f_i \geq 0, i \in 1, \ldots, |\mathbf{V}|$ the DAET of each actor v_i on any operator in $\mathbf{O}$ (Sect. 3.2),
- $l_j \geq 0, j \in 1, \ldots, P$ the load of each operator o_j, i.e. the DAET sum of the actors it executes,
- m_l the average of the operator loads.

$\mathbf{l} = [l_1, \ldots, l_P]$ is called the vector of loads and $\mathscr{L}$ the space of load repartitions with $\forall \mathbf{l}, \mathbf{l} \in \mathscr{L}$. The homogeneity of the architecture ensures the conservation of the work W:

$$W = \sum_{i=1}^{|\mathbf{V}|} f_i = \sum_{j=1}^{P} l_j. \tag{6.1}$$

`Problem 1: Latency Minimization.` Under the UNC conditions and with the constraint of minimizing latency, all actors in the critical path, and only these actors, should be assigned to a single operator that becomes the most loaded operator and contains no "hole" in its schedule. Otherwise, the number of operators increases without reducing the latency. The latency is then equal to the critical path and is thus minimal. In this case, the latency L is also equal to the highest load: $L = max_{j \leq P}(l_j)$. The procedure of minimizing the latency consists of searching for a load configuration $\mathbf{l}^*_{latency}$ so that:

$$\mathbf{l}^*_{latency} = \arg\min_{\mathbf{l} \in \mathscr{L}}(max_{j \leq P}(l_j)). \tag{6.2}$$

$max_{j \leq P}(l_j) \geq 1/P \sum_{i=1}^{|V|} f_i$ with equality if and only if a schedule exists with $l_j = W/P, \forall j$.

`Problem 2: Load Balancing.` To balance the loads, it is necessary to minimize the variance between the loads. If $m_l = W/P$ is the average of the loads and $\sigma^2 = 1/P \sum_{j=1}^{P}(l_j - m_l)^2$, minimizing the variance of the loads will balance the loads, and consists of searching for a load configuration $\mathbf{l}^*_{loads}$ such that:

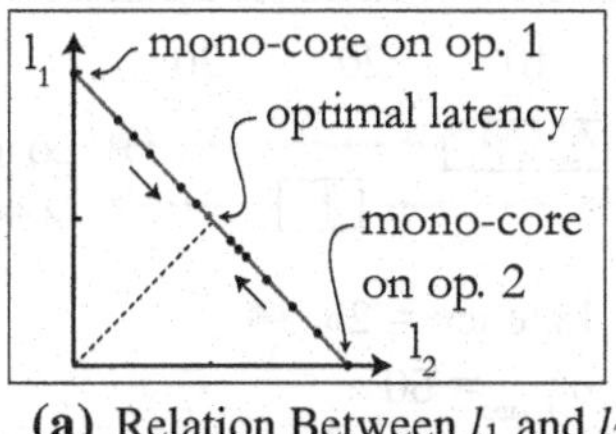

(a) Relation Between l_1 and l_2

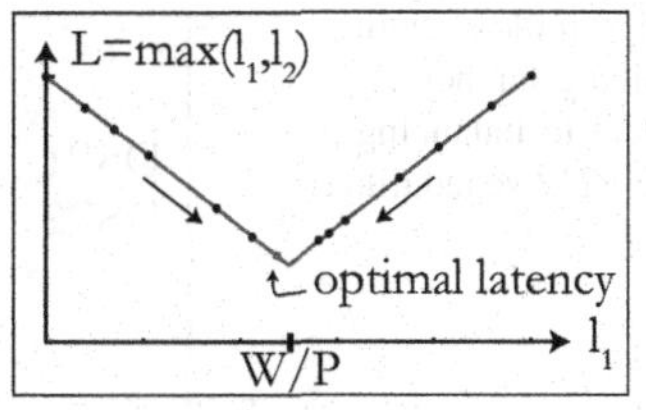

(b) Relation Between l_1 and L

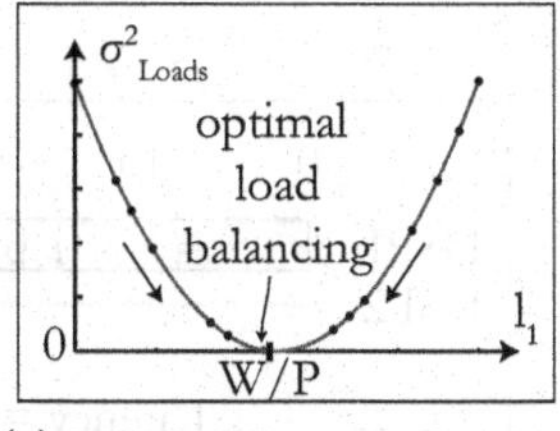

(c) Relation Between l_1 and σ^2

Fig. 6.9 Study of the latency and variance behavior in the case of 2 operators

$$\mathbf{l}^*_{loads} = \arg\min_{\mathbf{l}\in\mathscr{L}} \frac{1}{P}\sum_{j=1}^{P}(l_j - m_l)^2 = \arg\min_{\mathbf{l}\in\mathscr{L}} \frac{1}{P}\sum_{j=1}^{P} l_j^2. \tag{6.3}$$

Note that $\sigma^2 \geq 0$ and $\sigma^2 = 0$ if and only if the loads are perfectly balanced with $l_j = W/P, \forall j$. It may be hypothesized that any distribution of loads is possible:

Hypothesis 6.4.1 $\mathscr{L} = (\mathbf{R}^+)^P$.

The solutions for problems 1 and 2 are then identical and given by $l_j = W/P, \forall j$. In reality, the above Hypothesis 6.4.1 is not true because loads are partial sums of arbitrary DAETs f_i. The question then becomes: is the result still valid if $\mathscr{L} \neq (\mathbf{R}^+)^P$? The positive proof with $P = 2$ can be performed graphically.

In Fig. 6.9a, it may be noted that the discrete values of the loads of operators 1 and 2 fall on a line of best fit between the states which represent either o_1 or o_2 processing the whole work. The conservation of the work in Eq. 6.1 implies that l_2 depends on l_1 with $l_2 = W - l_1$. In Fig. 6.9b, the values of the latency $L = \max_{j\leq P}(l_j)$ are displayed relative to l_1.

In Fig. 6.9c, the variance between the loads σ^2 is shown to alter when l_1 changes. The two cost functions in Fig. 6.9b and c are convex with the same unique minimum point at $l_j = \dfrac{W}{P}, \forall j$. If l_1^* is the optimal load configuration for problem 1, it is also the optimal load configuration for problem 2. In the more general case of $P \geq 2$, the two functions are still convex functions due to the conservation of work. These two functions have the same unique minimum point corresponding to a latency $L = W/P$.

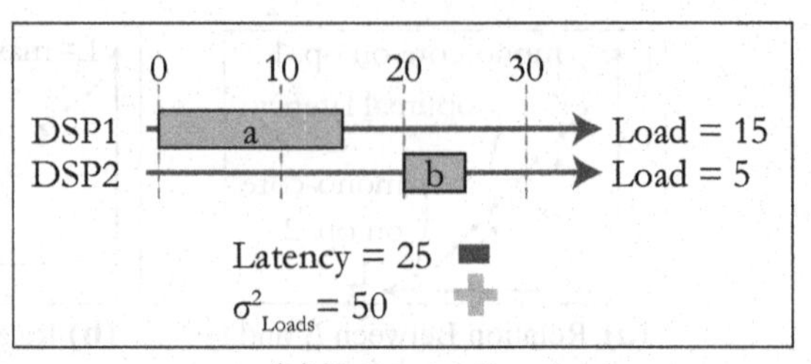

(a) Balancing Loads

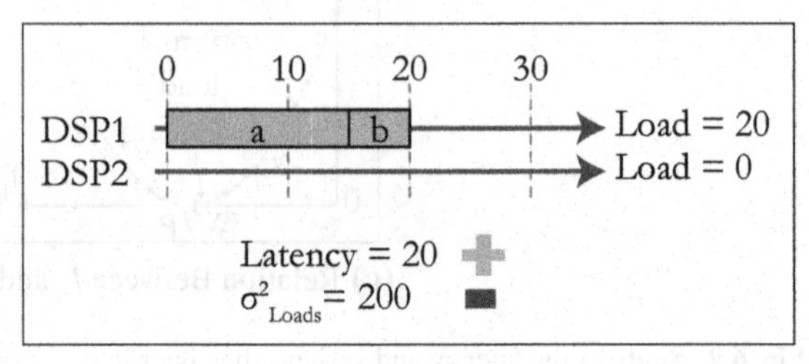

(b) Minimizing Latency

Fig. 6.10 Example showing that minimizing latency is not equivalent to balancing loads in the BNP scheduling problem

Therefore, it may be seen that the two problems are also equivalent in the case of discrete loads.

6.4.3.2 Non-Equivalence of Problems in BNP and APN Cases

Figure 6.10 illustrates a very simple example where minimizing the latency in a Bounded Number of Processor (BNP) scheduling (Sect. 4.4.3) makes the load balancing worse. The example of Fig. 6.10a, demonstrates that balanced loads result in a different assignment from that of Fig. 6.10b where latency is minimized. The success of load balancing is evaluated by the eventual variance of the loads σ^2_{Loads} that must be minimized. This evaluation procedure is explained below.

The two problems of latency minimization and load balancing are not equivalent in BNP conditions and even less so in APN conditions because transfer ordering is present which often penalizes distributed systems when minimizing latency.

6.4.3.3 Minimizing the Latency Under a Load Balancing Constraint

As the two problems of latency minimization and load balancing are not equivalent in the BNP case and even more divergent in the Arbitrary Processor Network (APN) case, the two constraints need to be considered separately during scheduling process and a good compromise between the two must be established. A basis for this tradeoff may be to employ a latency ABC that returns a composite cost which includes latency

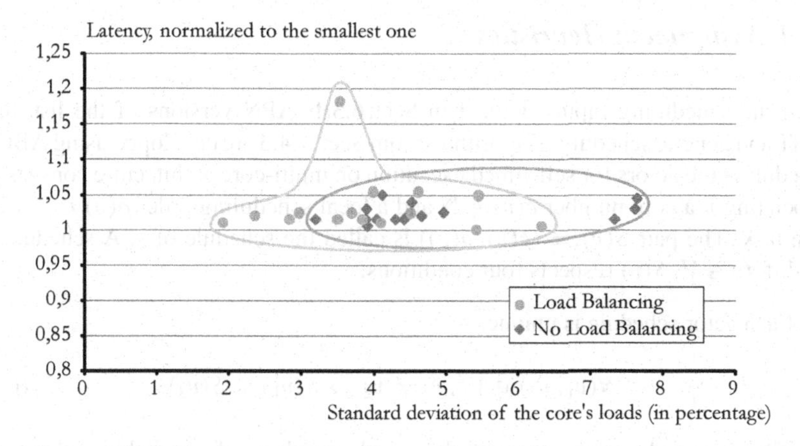

Fig. 6.11 Comparing schedules with and without load balancing criterium

minimization and load balancing:

$$C = L + \lambda.\sigma, \quad \text{with } \sigma = \sqrt{\frac{1}{N}\sum_{j=1}^{P}(l_j - m_l)^2}. \tag{6.4}$$

where L is the latency, λ a Lagrange multiplier and σ is the standard deviation of the loads. Experiments show that efficient results are obtained for a simple $\lambda = 1$. Figure 6.11 shows results of the rapid prototyping of a complete LTE description on the unnamed architecture presented in Sect. 5.1.1 and in [15]. The algorithm IBSDF after flattening has 464 vertices and 572 edges. The FAST algorithm is run 30 times, each time during 60 s with 5 s macrosteps. The loosely timed latency ABC is used. Points represent 15 solutions found with a composite cost and a Lagrange multiplier of 1, while diamonds represent solutions found minimizing only latency. It can be observed that the load balancing is significantly improved using the composite cost when, except for one abnormal solution, the latency is not significantly worsen.

6.5 Scheduling Heuristics in the Framework

There are two types of scheduling heuristics: assignment heuristics that assign actors to operators, and ordering heuristics that order actors on each operator and also order transfers on each Communication Node. Heuristics of both types are presented below As both these complementary types of scheduling heuristics are necessary for the solution, they are then combined in the scheduler structure as shown in Figs. 6.7 and 6.6.

6.5.1 Assignment Heuristics

From the scheduling inputs defined in Sect. 6.3.1, APN versions of the list, and FAST and genetic scheduling algorithms from Sect. 4.4.3 are developed using ABCs. Scheduling the actors for self-timed execution on multi-core architecture consists of associating v a core number $c(v) \in \mathbb{N}$ and a local scheduling order $o(v) \in \mathbb{N}$ with each task. The pair $S(v) = (c(v), o(v))$ is called the schedule of v. A schedule is valid if $\forall v \in V, S(v)$ respects four conditions:

1. Each actor schedule is unique:

$$\forall (v_1, v_2) \in V^2, v_1 \neq v_2 \Rightarrow S(v_1) \neq S(v_2). \tag{6.5}$$

2. The selected cores respect the total number of available operators: $c(v) < C, \forall v \in V$.
3. The local schedules respect the topological order of the graph: $\forall (v_1, v_2) \in V^2$ if there is a path from v_1 to v_2 and $c(v_1) = c(v_2)$, then $o(v_1) \leq o(v_2)$.
4. The implementation cost returned by the ABC respects: $cost(S(V)) < \infty$

6.5.1.1 Static List Scheduling

Static list scheduling using ABCs is described in Algorithm 6.1. An overview of this procedure was presented in Sect. 4.4.3. After the list ordering step, which is closely linked to infinite homogeneous simulation, the assignment heuristic can be used with any latency ABC. It is not straightforward to combine static list scheduling with the minimization of parameters other than latency because of the list ordering part and the necessity to evaluate partial schedules. However, it is possible to combine static list scheduling with the load balancing method presented in Sect. 6.4.3.

Algorithm 6.1: `Static_List_Scheduling`(G, M)

Input: A DAG $G = (V, E, w, c)$ and a route model M
Output: A schedule S of G on the route model M

1 Create an Infinite Homogeneous ABC: $IH - ABC(G, M)$ and a latency ABC: $L - ABC(G, M)$;
2 Retrieve the number of operators $|O|$ from $L - ABC(G, M)$;
3 $NodeList \leftarrow$ Sort the actors in V in CPN-Dominant order using costs from $IH - ABC(G, M)$;
4 **for** *each* $v \in NodeList$ **do**
5 $\quad i \leftarrow$ Select an operator index $o < |O|$ to execute v using costs from $L - ABC(G, M)$;
6 $\quad$ If no operator exists that can execute v with a finite cost, return an error;
7 $\quad$ Assign the actor v to o_i;
8 **end**

6.5.1.2 FAST and Genetic Heuristics

The underlying principles of the FAST algorithm from Kwok are discussed in Sect. 4.4.3. This algorithm uses an ABC and is described in Algorithm 6.2. The macro step and the FAST algorithm are halted when after a given amount of time. In PREESM, these times are specified in seconds and must be tuned to obtain good results. The programmer can stop the FAST process at any time, and the best schedule found is then returned. The time complexity is thus hard to evaluate and the randomness of the neighborood search makes the FAST algorithm non-deterministic.

Algorithm 6.2: `FAST_Scheduling`(G, S_0, M, $maxFastTime$, $maxMacrostepTime$)

Input: A DAG $G = (V, E, w, c)$, an initial schedule S_0, a route model M and two parameters $maxFastTime$ and $maxMacrostepTime$
Output: A schedule S of G on the route model M

1 Set $S = S_0$;
2 Create an ABC of desired type $ABC(G, M)$;
3 reset fastTime;
4 **while** $fastTime < maxFastTime$ *and the programmer did not stop the process* **do**
5   reset macrostepTime;
6   Change the assignment of a randomly chosen Critical Path Node (CPN) v_{CPN};
7     $S_{store} = S$;
8   **while** $macrostepTime < maxMacrostepTime$ **do**
9   Change the assignment of a randomly chosen non-CPN node $v_{non-CPN}$;
10   **if** ABC *returns a higher cost than before* **then**
11   Change back the assignment;
12   **end**
13   **end**
14   **if** ABC *returns a higher cost than before* **then**
15         $S = S_{store}$;
16   **end**
17 **end**
18 **return** S;

The fact that the FAST algorithm only assesses the cost of completely scheduled applications makes it compatible with any type of ABC. It could thus be used to minimize memory, power consumption or a joint cost function. Keeping the best outputs of the macro steps, a population of schedules can be generated and mutations and cross-overs can be applied recursively (Sect. 4.4.3) in a genetic algorithm. A genetic heuristic is a means of improving schedules without being blocked in a locally optimal point. It can also be combined with any type of ABC.

6.5.2 Ordering Heuristics

In the scheduling algorithms presented above, the ABC recalculates the order of an actor when the heuristic assigns an actor to a new operator. Moreover, when transfer competition is taken into account on a Contention Node (CN), the order of the transfer is also computed. The algorithm used to recalculate this order can be changed to obtain a better compromise between scheduling time and schedule quality.

Currently, two algorithms are implemented in PREESM:

- The `simple ordering heuristic` follows the CPN-Dominant scheduling order given by the task list resulting from the list scheduling algorithm. Transfers are assigned an order related to their sender or receiver. This order is likely to be quite sub-optimal but there is a fixed complexity O(1).
- The `switching ordering heuristic` is more accurate. When a new actor or transfer needs to be scheduled, the algorithm looks for the earliest `hole` in the operator or CN schedule of sufficient size to contain the candidate. This hole must be after the candidate's predecessors in the schedule so as not to introduce new costly synchronizations. It then inserts the actor or transfer in this hole. This algorithm is in $O(|V|(|E|+|V|))$, which can greatly increase the general complexity because it is applied each time an actor is assigned or a transfer is scheduled. However, the algorithm execution time is still realistic in practice and is counterbalanced by good performance.

The scheduler framework enables the comparison of different edge scheduling algorithms using the same task scheduling sub-module and architecture model description. The switching ordering heuristic has good performance when combined with FAST on a system with transfer contention because it adaptively reschedules actors and transfers during the macro and micro steps (Algorithm 6.2).

The previous section discussed methods for obtaining good schedules. Next, the method of assessing the quality of a schedule will be explained.

6.6 Quality Assessment of a Multi-Core Schedule

Quality assessment of an automatically generated schedule is an important feature of a multi-core development chain because it identifies the weaknesses of the system. This section details a graphical quality assessment chart which pinpoints a lack of algorithm or architecture parallelism, in addition to diagnosing underperforming schedules.

6.6.1 Limits in Algorithm Middle-Grain Parallelism

The obvious metric for algorithm parallelism in terms of latency is the speedup $S(n) = T(1)/T(n)$ where $T(1)$ is the latency on one core, also called the `work` (Sect. 6.4.2) and $T(n)$ the latency on n cores. Other metrics of parallelism based on execution time (Sect. 6.1.1) could be developed but in the case of LTE, latency is the primary time constraint. Maximal algorithm speedups started to be discussed in 1967, when Amdahl describes an evaluation method of an intrinsic algorithm parallelism. His theorem simply states that if a portion r_s of a program is sequential, the execution speedup that may be attained using multiple homogeneous cores compared to one core is limited to $1/r_s$, independent of the number of cores [16]. For example, a program with 20 % of sequential code has a speedup limit of 5. This formula, known as the Amdahl's law, subsequently led to pessimism about the future of multi-core programming. Indeed, if as low as 20 % of sequential code limits the speedup to five, creating architectures with more than a few cores may be useless. Figure 6.12a illustrates this limit for several cases of sequential program portions. The maximal speedup according to Amdahl's law on n homogeneous cores is given by:

$$S \leq S_{max_{Amdahl}}(n) = \frac{1}{r_s + \dfrac{1 - r_s}{n}}. \tag{6.6}$$

Thankfully, this formula is an over-simplified view of reality, as it defines a perfectly parallel fixed program portion that can be distributed over any number of cores and a fixed sequential program portion, where execution time does not depend on the number of executing cores. These asumptions are based on the idea that a main execution thread spawns parallel threads when data-parallel operations need to be executed and retrieves the data after the end of the parallel program portion. Such model is far from exploiting all the parallelism from an application (data parallelism as well as pipelining). In [17], Gustafson proposes a new model where the serial section of the code represents the same portion of the execution, regardless of the number of cores. The inherent idea is that more cores enable the execution of new portions of the code in parallel. For example, the execution on a given number of cores will spend 20 % of the time on sequential code. The execution time on one core is then $T(1) = r_s(n) + n(1 - r_s(n))$ for any n when the execution time on n cores is $T(n) = r_s(n) + (1 - r_s(n)) = 1$. The resulting maximal speedup on n homogeneous cores is shown in Fig. 6.12b for several sequential part ratios. It is called Gustafson-Barsis law and given by:

$$S \leq S_{max_{Gustafson}}(n) = n - r_s.(n - 1). \tag{6.7}$$

While Amdahl's law is pessimistic for the parallel calculation possibilities of an algorithm, the Gustafson-Barsis law is optimistic and shows that Amdahl's law was not a strict limit, as previously believed. With a sequential section of 20 %, the Amdahl maximal speedup on 10 cores is 3.6 while the Gustafson maximal speedup is

Fig. 6.12 Theoretical speedup limits

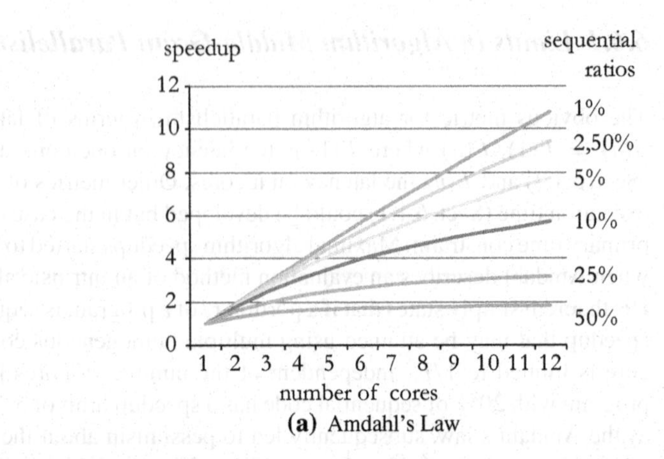

(a) Amdahl's Law

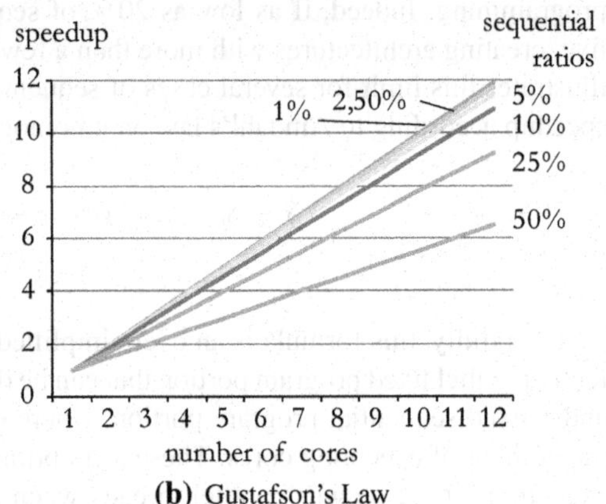

(b) Gustafson's Law

8.2. Amdahl and Gustafson laws, as well as Karp-Flatt metric [18], are based on the very unrealistic idea that the middle-grain parallelism of algorithm is concentrated in perfectly parallel zones. With graphs of program execution, more realistic parallelism metrics can be used as advised by Leiserson in [19]. Contrary to the previous laws, Leiserson's intent is only to evaluate the available parallelism present in the high-level model of the code; however, the figures are more reliable because they are based on algorithm properties that are more precise than parallel and sequential program sections.

6.6.2 Upper Bound of the Algorithm Speedup

It is important to evaluate the quality of the computed schedule. For that purpose, the speedup increase that occurs as the number of cores is augmented is of interest. For this study, communication times are not considered because they greatly depend on the architecture. The only input for the evaluation of the possible speedup included in this study is the algorithm. It is assumed that the architecture is homogeneous and has an infinite speed media. The first limit that must be considered is the work length. The work is the sum of all actor execution times and corresponds to the algorithm latency on one core T(1) (Sect. 6.4.2). Except in the exotic case of superlinear speedups which appear when the work decreases with the increasing number of cores, the work will be distributed on all n cores and so:

$$T(n) \geq T(1)/n \Rightarrow S \leq S_{sup_{work}} = n. \tag{6.8}$$

This speedup limit is a function of the number of available cores. Another general limitation can be found in the length of the `span`, i.e. the critical path when communications are ignored. Both work and span were shown to be calculable by the infinite homogeneous ABC in Sect. 6.4.2. No matter how well the algorithm is parallelized, the latency can not be reduced to less than the critical path. The critical path length T_∞ adds a new constraint on the speedup:

$$T(n) \geq T_\infty \Rightarrow S \leq S_{sup_{span}} = T(1)/T_\infty. \tag{6.9}$$

As these upper bounds have been defined, a minimum acceptable performance may be researched.

6.6.3 Lowest Acceptable Speedup Evaluation

As compile-time scheduling computation is used with list scheduling and probabilistic refinements, its performance should always be equal to or better than the basic "greedy" scheduler. The "greedy" scheduler executes actors as soon as they become available. At each time step, there are either more than n available actors, where n of them are dispatched on the cores and the step is called `complete` or there are less than n available actors, and they are all dispatched on the cores and the step is called `incomplete`. Graham [20] and Brent [21] give an limit inferior to the latency that such a "greedy" scheduler can obtain using the Greedy Scheduling Theorem (GST):

$$T(n) \leq \frac{T(1)}{n} + T_\infty \Rightarrow S \geq S_{inf_{GST}} = \frac{T(1)}{\frac{T(1)}{n} + T_\infty}. \tag{6.10}$$

The proof of GST can be found in [19]. If a result worse than GST is obtained, the compile-time scheduling process can not be considered successful. Again, this theorem does not take into account the communication times due to hardware constraints that can make real schedules sub-optimal. From the previously defined upper

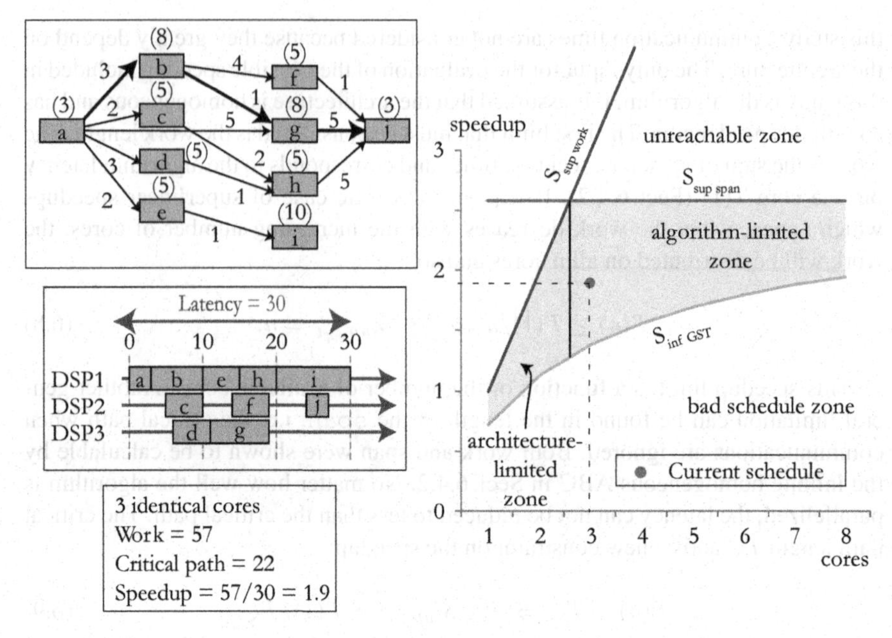

Fig. 6.13 Speedup chart of an example DAG scheduling

and lower bounds on speedup, a chart can be derived, displaying the optimality of an algorithm schedule latency.

Figure 6.13 displays the speedup optimality assessment of a small DAG scheduling procedure. The DAG is annotated with its Deterministic Actors Execution Times (in brackets) and the time cost of its transfers. To be efficient, the scheduling must be inside the space delimited by the three curves. The best results are obtained when it is closest to the upper left corner. Moreover, depending on the number of cores, its parallelism is either limited by the number of cores in the architecture (architecture-limited zone) or by the degree of parallelism in the graph (algorithm-limited zone). In [22], Sinnen defines the Communication to Computation Ratio (CCR) as the sum of communication time costs divided by the sum of computation time costs in a scheduled graph. Figure 6.13 makes sense as long as $CCR \ll 1$ is in the tested schedule. This constraint is not a real problem because $CCR \ll 1$ is also a condition for the implementation to be efficient.

6.6.4 Applying Scheduling Quality Assessment to Heterogeneous Target Architectures

The previous calculations were all based on homogeneous architectures but their use can be extended to heterogeneous target architectures if certain precautions are taken. The target architectures of this study (Sect. 5.1.1) have cores of homogeneous type c64x+ that perform the majority of the operations in the application. The c64x+ core type must be set in the scenario as the main operator type. The parallelism of the cores with main type can thus be studied in the quality assessment chart. The results are complicated by the fact that coprocessors improve the speedup and increased communications degrade the speedup. The number of cores in the speedup chart refer only to cores with main operator type. The second precaution is that any actor executed by coprocessors must be forced on that coprocessor to allow the specific study of the behavior of c64x+ cores.

In order to compute work and span, the infinite homogeneous ABC is run to simulate a UNC execution of the code (Sect. 6.4.2). It calculates the algorithm work using DAETs with software actors on maintype cores and hardware actors on their respective coprocessors. The span is different than the critical path length computed for the list scheduling because it does not take into account the transfers in the infinite homogeneous architecture. The span is computed by scheduling the graph on a UNC architecture with an infinite number of main type cores and coprocessors with non-null loads.

Under these conditions, the speedup chart provides a quick evaluation of the optimality of the main system cores under the constraint of fixed mappings. It can be used to evaluate whether a lower granularity in the algorithm is necessary to extract more parallelism or, conversely, if certain actors should be clustered.

After this introduction of enhanced rapid prototyping features, the following chapters will discuss the modeling of LTE for rapid prototyping and code generation.

References

1. Gatherer A, Biscondi E (2009) Multicore DSP programming models [In the spotlight]. IEEE Signal Process Mag 26(6):224, 220–222. doi:10.1109/MSP.2009.934182
2. Karam LJ, AlKamal I, Gatherer A, Frantz GA, Anderson DV, Evans BL (2009) Trends in multicore DSP platforms. IEEE Signal Process Mag 26(6):38–49
3. Haid W, Huang K, Bacivarov I, Thiele L (2009) Multiprocessor SoC software design flows. IEEE Signal Process Mag 26(6):64–71
4. Piat J, Bhattacharyya SS, Pelcat M, Raulet M (2009) Multi-corecode generation from interface based hierarchy. DASIP 2009
5. Pino JL, Lee EA (1995) Hierarchical static scheduling of dataflow graphs onto multiple processors. IEEE Int. Conf. Acoust. Speech Signal Process. 4:2643–2646. doi:10.1.1.17.8262. http://citeseerx.ist.psu.edu/viewdoc/summary?doi=10.1.1.17.8262
6. Piat J (2010) Data flow modeling and multi-core optimization of loop patterns. Ph.D. thesis, INSA Rennes
7. Grandpierre T, Sorel Y (2003) From algorithm and architecture specifications to automatic generation of distributed real-time executives: a seamless flow of graphs transformations. In: MEMOCODE '03, pp 123–132
8. Graphiti Editor : Available. http://sourceforge.net/projects/graphiti-editor/
9. Pelcat M, Menuet P, Aridhi S, Nezan J (2009) Scalable Compile-Time scheduler for multi-core architectures. In: DATE, 2009
10. w3c XSLT standard. http://www.w3.org/Style/XSL/
11. Open SystemC initiative web site. http://www.systemc.org/home/
12. Ghenassia F (2006) Transaction-level modeling with systemC: TLM concepts and applications for embedded systems. Springer, New York.http://portal.acm.org/citation.cfm?id=1213675

13. TMS320TCI6488 DSP platform, texas instrument product bulletin (sprt415) (2007)
14. Pelcat M, Nezan JF, Piat J, Croizer J, Aridhi S (2009) A system-level architecture model for rapid prototyping of heterogeneous multicore embedded systems. In: DASIP. http://hal.archives-ouvertes.fr/hal-00429397/en/
15. Friedmann A (2010) Enabling LTE development with TI new multicore SoC architecture SPRY134. Technical report, Texas Instruments
16. Amdahl GM (1967) Validity of the single processor approach to achieving large scale computing capabilities. In: Proceedings of the April 18–20, 1967, spring joint computer conference, pp 483–485
17. Gustafson JL (1988) Reevaluating amdahl's law. Commun ACM 31(5):532–533
18. Karp AH, Flatt HP (1990) Measuring parallel processor performance. Commun ACM 33(5):539–543
19. Leiserson CE (2005) A minicourse on dynamic multithreaded algorithms. MIT Press, Cambridge
20. Graham RL (1969) Bounds on multiprocessing timing anomalies. SIAM J Appl Math 17: 416–429. doi:10.1.1.90.8131. http://citeseerx.ist.psu.edu/viewdoc/summary?
21. Brent RP (1974) The parallel evaluation of general arithmetic expressions. J ACM 21(2): 201–206
22. Sinnen O (2007) Task scheduling for parallel systems (Wiley series on parallel and distributed computing). Wiley-Interscience, Hoboken

Chapter 7
Dataflow LTE Models

7.1 Introduction

The objectives of rapid prototyping are introduced in Chap. 1. Figure 6.1 illustrates the process of rapid prototyping. Technical background on the subject is explored in Chap. 4. In this chapter, models for the LTE rapid prototyping process are explained. From these models, execution can be simulated and optimized using multi-core scheduling heuristics and code can also be generated. The LTE models are novel and can complement the standard documents for a better understanding of the LTE eNodeB physical layer. After a general view of the LTE model is given in Sect. 7.2, the three parts of the LTE eNodeB physical layer are detailed in Sects. 7.3–7.5. LTE rapid prototyping is processed by a Java-base framework which includes PREESM. The elements of this framework are introduced in following sections.

7.1.1 Elements of the Rapid Prototyping Framework

The framework for rapid prototyping that was constructed contains three main elements: PREESM, Graphiti and SDF4J. The framework is illustrated in Fig. 7.1 and detailed in [1]. All the blocks in Fig. 7.1 except SDF4J are open-source plug-ins for the Eclipse environment [2]. They can use and extend the advanced functionalities of Eclipse in terms of graphical user interface and framework organization. In the PREESM project, special attention is given to program tools that will be maintained in the long term, using encapsulation and design patterns [3] (visitor, abstract factory, singleton, command…). Execution time of the different workflow elements is optimized to offer prototyping solutions as expediously as possible.

M. Pelcat et al., *Physical Layer Multi-Core Prototyping*,
Lecture Notes in Electrical Engineering 171, DOI: 10.1007/978-1-4471-4210-2_7,

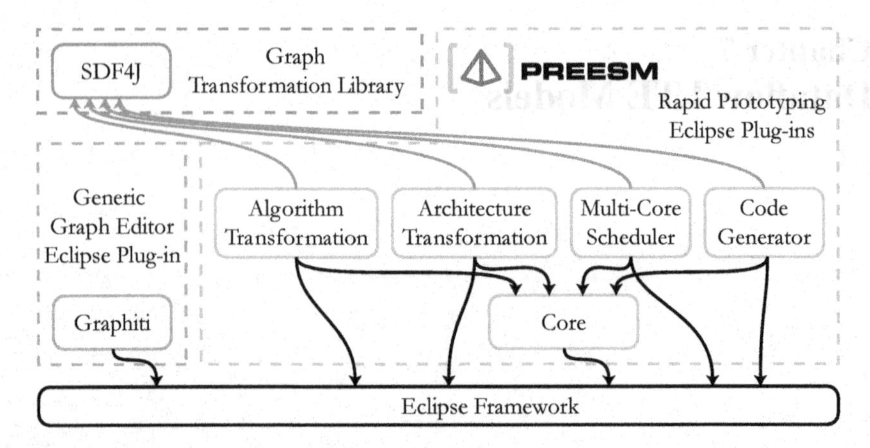

Fig. 7.1 An eclipse-based rapid prototyping framework

7.1.2 SDF4J: A Java Library for Algorithm Graph Transformations

The java library named SDF4J (Synchronous Dataflow for Java) manipulates dataflow graphs and is available in Sourceforge [4]. This library can process SDF graphs (Sect. 3.2) and its subsets (DAG...) as well as IBSDF (Sect. 3.5.2). It performs the algorithm graph transformations called in the workflows, as explained in Sect. 6.2.3, including hierarchy flattening, schedulability verification (Sect. 3.2), transformation into a DAG (Sect. 3.2.3), and clustering methods based on [5]. SDF4J also contains a parser and a writer of GraphML files [6] to load and store its graphs.

7.1.3 Graphiti: A Generic Graph Editor for Editing Architectures, Algorithms and Workflows

Graphiti provides a generic graph editor and is completely independent from PREESM. It is written using the Graphical Editor Framework (GEF). The editor is generic in the sense that any type of graph may be represented and edited. Graphiti is routinely used with the following graph types and associated file formats : CAL networks [7, 8], S-LAM architectural representations stored in IP-XACT format (Sect. 5.2), SDF and IBSDF graphs stored in GraphML format and PREESM workflows (Sect. 6.2.3), stored in specific XML files.

The type of graph is registered within the editor by a `configuration`. A configuration is a structure that describes:

1. The `abstract syntax` of the graph (types of vertices and edges, and attributes allowed for objects of each type).

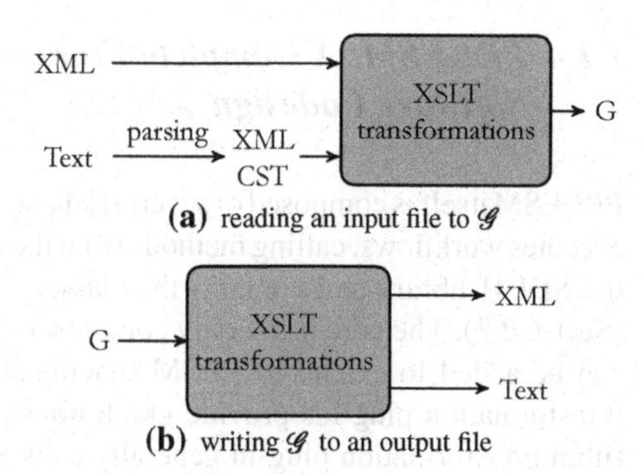

Fig. 7.2 Input/output with Graphiti's XML format $\mathcal{G}$

2. The `visual syntax` of the graph (colors, shapes, and so on).
3. Transformations from the file format in which the graph is defined to Graphiti's XML file format $\mathcal{G}$, and vice versa (Fig. 7.2).

Two kinds of input transformations are supported: from custom XML to Graphiti XML format and from text to Graphiti XML format (Fig. 7.2). XML is transformed to XML using Extensible Stylesheet Language Transformation (XSLT [9]). The second input transformation parses the input text to its Concrete Syntax Tree (CST) represented in XML according to a LL(k) grammar by the Grammatica [10] parser. Similarly, two kinds of output transformations are supported, from XML to XML and from XML to text.

Graphiti handles `attributed graphs` [11]. An attributed graph is defined as a directed multigraph $G = (V, E, \mu)$ where V the set of vertices, and E the multiset of edges (there can be more than one edge between any two vertices). μ is a function $\mu : (\{G\} \cup V \cup E) \times A \mapsto U$ that associates instances with attributes from the attribute name set A and values from U, the set of possible attribute values. A built-in `type` attribute is defined so that each instance $i \in \{G\} \cup V \cup E$ has a type $t = \mu(i, \texttt{type})$, and only admits attributes from a set $A_t \subset A$ given by $A_t = \tau(t)$. Additionally, a type t has a visual syntax $\sigma(t)$ that defines its color, shape and size.

To edit a graph, the user selects a file and the matching configuration based on the file extension is computed. The transformations defined in the configuration file are then applied to the input file and result in a graph defined in Graphiti's XML format $\mathcal{G}$ as shown in Fig. 7.2. The editor uses the visual syntax defined by σ in the configuration to draw the graph, vertices and edges. For each instance of type t the user can edit the relevant attributes allowed by $\tau(t)$ as defined in the configuration. Saving a graph consists of writing the graph in $\mathcal{G}$, and transforming it back to the input file's native format.

Graphiti generates workflow graphs, IP-XACT and GraphML files that are the primary inputs of PREESM rapid prototyping. The GraphML files containing the algorithm model are loaded and stored in PREESM by the SDF4J library.

7.1.4 PREESM: A Complete Framework for Hardware and Software Codesign

PREESM itself is composed of several Eclipse plug-ins (Fig. 7.1). The core parses and executes workflows, calling methods from the other plug-ins of PREESM. It includes the SDF4J library and contains the classes of all objects exchanged in workflows (Sect. 6.2.3). The core is the only compulsory element for PREESM. Other plug-ins can be added to extend PREESM functionalities. The algorithm and architecture transformation plug-ins provide graph transformations to the workflow. The algorithm transformation plug-in generally calls SDF4J functionalities. The multi-core scheduler contains all the features discussed in Chap. 6. The code generation plug-in generates self-timed executable code (Sect. 4.4.1) from schedules provided by the multi-core scheduler. Code generation will be extended in Chap. 8.

7.2 Proposed LTE Models

7.2.1 Fixed and Variable eNodeB Parameters

Certain parameters of an eNodeB do not change during its whole life span and other parameters change every millisecond. For the rapid prototyping process, simulation use cases must be chosen that represent real-life execution cases. The compile-time and run-time actor scheduling must take into account the life spans of parameters and to manage static parameters, no scheduling overhead should be introduced at run-time.

The `duplex mode` (Sect. 2.3.4) is a fixed parameter determined during the network construction. The `bandwidth` of an eNodeB and the cyclic prefix mode (normal or extended) are also fixed network parameters. Once a frequency band has been bought by an operator, the eNodeB is configured to exploit fully this bandwidth. The number of available `subcarriers` is consequently also a fixed parameter, as are the `size of the Fourier transforms` in SC-FDMA and OFDMA encoding and decoding (Sect. 2.3.4).

The format and localization of the RACH preamble, as well as the number of roots and cyclic shifts (Sect. 2.4.5) are stable. These parameters depend only on the environment of the eNodeB and are not permitted to vary over time because they must provide a stable access point for UEs. Thus, the RACH decoding has no variable parameters.

The `number of antennas` for uplink, downlink and PRACH is stable. The `multiple antenna schemes` for uplink and downlink are determined in the eNodeB (Sects. 2.3.6, 2.4.4 and 2.5.4). Each eNodeB may have a preferred multiple antenna scheme corresponding to its environment (rural and flat, mountainous, urban…) but a number of different modes are likely to be used depending on the channel quality of each user.

It is the highly variable parameters of an eNodeB that depend on the UE connections. In PUSCH, the `number of UEs` sending data changes every TTI of one millisecond (Sect. 2.3.4). The `number of Code Blocks` sent (Sect. 2.3.5) and the `sizes of these Code Blocks` also changes every TTI depending on the services accessed by each UE (telephony, web...). The repartition of this data in the frequency band can differ for each slot of 0.5 ms because of frequency hopping (Sect. 2.4.2). In PDSCH, the number of variables is approximately the same as in PUSCH but with an additional degree of freedom in the `number of transport blocks sent to a UE` during a TTI. In PUSCH, with the exception of MU-MIMO, there is no spatial multiplexing so only one transport block per TTI is transmitted (Sect. 2.4.4). In PDSCH, 2×2 spatial multiplexing can be employed (Sect. 2.5.4) and a single UE can receive two transport blocks in one TTI.

The `modulation and coding scheme` (Sect. 2.3.5) is specific to each UE so will also varies in every TTI. The transport block size and number of transport blocks for a UE determines for each TTI the quantity of data exchanged with the core network (Fig. 2.4).

As a consequence of these parameter variations, RACH preamble decoding is a static operation that can be studied entirely at compile-time while uplink decoding and downlink encoding are highly variable over time. Despite these variations, case studies may be chosen so behavioral information can be extracted from the corresponding dataflow graphs at compile-time. A typical use case is presented in next section.

7.2.2 A LTE eNodeB Use Case

In order to study LTE multi-core execution, typical use cases must be defined. We present here a typical eNodeB and its performance. The objective of this section is to explain the important features that must be taken into account when evaluating LTE physical layer performance. Considering a FDD eNodeB with 20 MHz downlink and uplink and 100 PRBs per slot, the system behavior can be evaluated using information given in Chap. 2. A configuration of four transmission and four reception antenna ports (Sect. 2.3.6) is chosen for this eNodeB.

7.2.2.1 Uplink Performances

Firstly, considering the PUSCH with standard cyclic prefix, each PUSCH resource block contains $6\,symbols * 12\,subcarriers = 72\,resourceelements$; one additional symbol is used as a reference symbol (Fig. 2.17). Table 7.1 gives the capacity of Resource Elements (RE) and resource blocks in bits and deduces the raw bit rates for the current case. This total data rate must be shared between all UEs. The base time unit for data allocation to a UE is a single TTI of one millisecond; thus it can be seen that a pair of PRBs contain the minimum amount of bits that can be allocated to a UE.

Table 7.1 Maximal raw bit rates of uplink resources

Modulation scheme	bits/RE	bits/PRB	bits/pair of PRBs	max raw bitrate (Mbit/s)
QPSK	2	144	288	28.8
16-QAM	4	288	576	57.6
64-QAM	6	432	864	86.4

From these raw data rates, the multiple control overheads must be subtracted:

- The outer PRBs are reserved for PUCCH. This exact number is typically up to 16 PUCCH regions, i.e. 16 PRBs per slot are kept for PUCCH [12]. It may be noted that the modulation and coding scheme is different between PUSCH and PUCCH; thus the PUCCH raw bit rate cannot be directly deduced from this calculation.
- The rate matching process (Sect. 2.3.5) has a rate of "useful data" between 7.6 and 93 %. The channel coding rate is linked to the chosen modulation scheme (Fig. 2.14a).
- The transmission of the PRACH channel requires part of the uplink bandwidth. In a typical case of a cell with radius smaller than 14 km, a preamble of format 0 can be allocated every 2 ms . It results in 6 PRBs every 2 ms, so uses approximately 3 % of the resources.

Thus, the maximal PUSCH data rate may be evaluated, including these overheads. Under ideal conditions where 64-QAM can be employed over the entire bandwidth with 93 % channel coding rate, a PUSCH data rate of $86.4 \times 0.93 \times 0.81 = 65$ Mbit/s can be attained. The CRCs and the overhead of the upper layers are not taken into account; these parameter also reduce the final bit rate available to the UEs IP layer. The uplink transmission process is constructed to allow the LTE uplink data rate requirement of 50 Mbit/s to be attained.

7.2.2.2 Downlink Performances

The downlink performance evaluation uses the table of transport block sizes in [13] p. 26. The transport block size (in bits) depends on the number of allocated PRBs, and on an index named I_{TBS} (Index of Transport Block Size). Downlink Control Information (DCI, Sects. 2.3.5 and 2.5.2) sent in PDCCH to a UE details the link adaptation parameters and determines the I_{TBS}. From the I_{TBS}, the maximal downlink data rates for a given modulation scheme, with 2×2 spatial multiplexing (2 transport blocks) can be evaluated. The results are shown in Table 7.2.

The displayed data rates target only PDSCH and include the control channel costs. As above, these maximal data rates are those of the physical layer. Upper layers will further reduce the effective data rate available to the IP layer of the UEs.

The following sections focus on models of multi-core execution for the physical layer of LTE eNodeBs. To study the system-level and the link-level data behavior

Table 7.2 Maximal raw bit rates of PDSCH

Modulation scheme	I_{TBS}	bits/2TBs	max raw bitrate (Mbit/s)
QPSK	9	31,704	31.7
16-QAM	15	61,664	61.7
64-QAM	26	1,49,776	149.8

of the LTE downlink, an open source simulator under Matlab is provided by the Technical University of Vienna [14].

7.2.3 The Different Parts of the LTE Physical Layer Model

LTE physical layer decoding can be divided into three major actors: `random access decoding`, `uplink decoding` and `downlink encoding`. Uplink decoding and random access decoding are frequency multiplexed in the same symbols but they are separated in this study for two reasons: firstly, both actors are very costly, and are potentially parallel and secondly, while PRACH decoding is a completely static operation, uplink decoding execution varies every millisecond. Thus studying each actor separately allows better and fast optimization.

The scale and granularity of the three actor descriptions must be defined (Sect. 3.1.2). The scale is determined by the application. For PRACH decoding, the obvious scale results from a graph decoding a complete preamble sent over 1, 2 or 3 ms. This preamble transmission time depends on preamble type. The latency of one RACH preamble detection then is minimized. For uplink and downlink actors, graphs could be constructed where each processes one symbol (decoding a 71.4 μs period) but this would introduce too much low-level conditioning, including parameters with static patterns.

Another possibility would be to design uplink and downlink graphs representing one entire frame (decoding 10 ms) but this solution leads to very large graphs and HARQ puts latency constraint on subframe processing, not frame processing. Thus, the correct scale for uplink and downlink graphs is the time to process one subframe (1 ms). The granularity is chosen so that atomic actors (with no hierarchy) have relatively close execution times. A typical actor execution time is a few thousand of cycles. In the following sections, models of random access decoding, uplink decoding and downlink encoding with user- selected scale will be presented.

7.3 Prototyping RACH Preamble Detection

Random Access Channel Preamble Detection (RACH-PD) consists of decoding the multiplexed and non-synchronized messages from the UEs attempting to connect to the eNodeB (Sect. 2.4.5). The preamble is transmitted on the Physical RACH

(PRACH) channel over a specified time–frequency resource, denoted as a `slot`, available with a certain cycle period and a fixed bandwidth of 6 PRBs. Within each slot, a guard period is reserved at each end to maintain time orthogonality between adjacent slots.

The method used to execute RACH-PD is the hybrid time–frequency domain approach described in [12] p. 449 and [15]. With this method, the PRACH message decoder can start before the whole message has been received. It uses small FFTs and is based on downsampling and anti-aliasing. The Power Delay Profile (PDP) is then computed. It consists of the norm of the periodic correlation of each Zadoff–Chu sequence (Sect. 2.4.3) with the received sequence. A peak in the Power Delay Profile indicates the detection of a signature and the location of the peak provides the timing advance of the transmitting UE, i.e. the propagation time between UE and eNodeB. Peak detection must take into account the noise in received samples; thus, the noise is estimated and the threshold for peak detection depends on this estimation. To compute the correlation between received signal and signatures, the convolution with complex conjugate is used.

The case studied in this section assumes a RACH-PD for a cell size of 115 km. This is the largest cell size supported by LTE and is also the case requiring the most processing power. According to [16], preamble format #3 is used with 21,012 complex samples as a cyclic prefix, followed by a preamble of 24,576 samples followed by the same 24,576 samples repeated. In this case, the slot duration is 3 ms which gives a guard period of 21,996 samples.

As per Fig. 7.3, the algorithm for the RACH preamble detection can be summarized in the following steps [15]. The dataflow graph in Fig. 7.3 illustrates the actors contained by the IBSDF description.

1. After cyclic prefix removal, the `pre-processing actor` isolates the RACH bandwidth by shifting the data in frequency (Frequency Shifting, FS) and band-pass filtering it with downsampling (Finite Impulse Response, FIR). The data is then transformed into the frequency domain (DFT) and the subcarriers of interest are selected (subcarrier demapping).
2. Next, the `circular correlation actor` correlates data with pre-stored preamble root sequences in order to discriminate between simultaneous messages from several users. The complex conjugates of the root sequences are directly retrieved from memory; consequently, the correlation step only costs a complex multiplication. The circular correlation actor also applies an IFFT to return to the temporal domain and calculates the energy of each root sequence correlation. The more roots there are, the more circular correlations are processed. This is why cyclic shifts are used when possible to reduce the number of root (Sect. 2.4.5).
3. Then, the `noisefloor threshold actor` collects these energies and estimates the noise level for each root sequence.
4. Finally, the `peak search actor` detects all signatures sent by the users in the current time window. It additionally evaluates the transmission timing advances, correlated with the distance of the transmitting UEs. The timing advances are

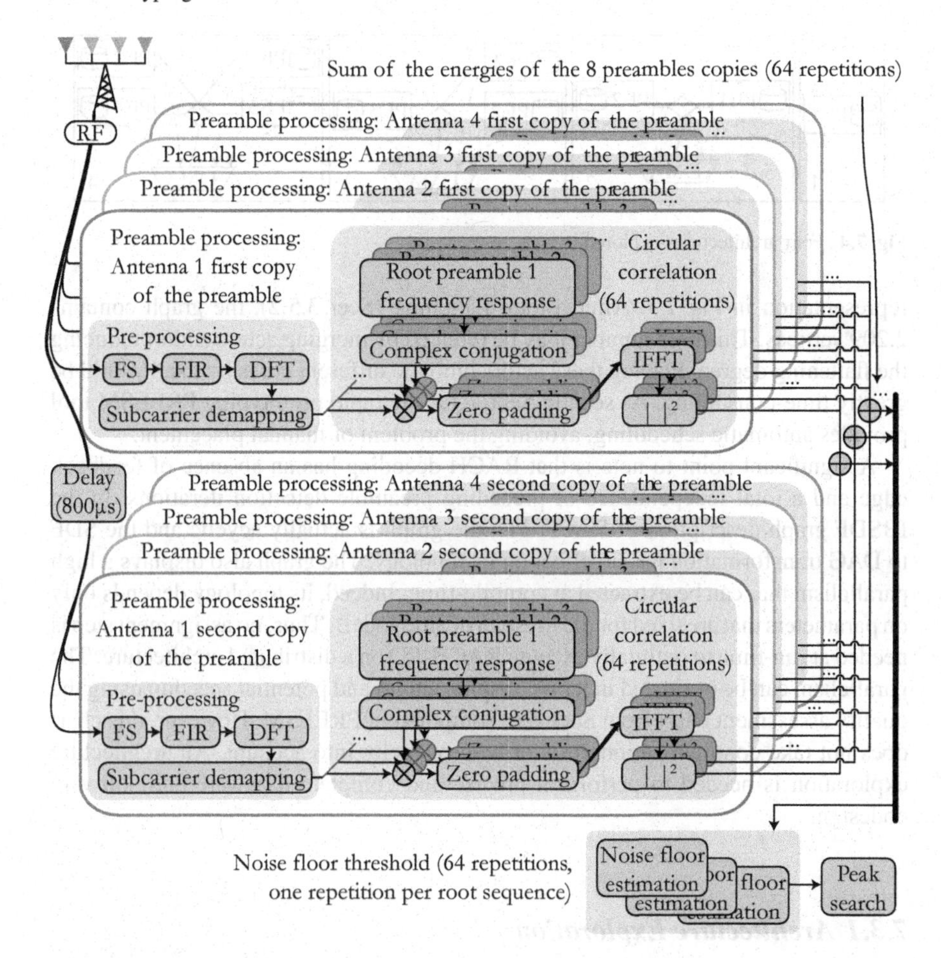

Fig. 7.3 RACH-PD algorithm model

sent to the MAC layer which then transmits each value to the appropriate UE to command a compensation.

In general, there are three parameters of RACH which are highly dependent on the cell size, and may be varied: the number of receive antennas, the number of root sequences and the number of repetitions of the same preamble. The case with 115 km cell size implies four antennas, 64 root sequences, and two repetitions, as shown in Fig. 7.3.

The goal of rapid prototyping of a RACH-PD can be to determine, through simulation, the number of c64x+ cores needed by the architecture to manage the 115 km cell RACH-PD algorithm. The RACH-PD algorithm behavior is described as a IBSDF graph in PREESM. The algorithm can be easily adapted to different eNodeB use cases by tuning the graph parameters. The IBSDF description is derived from the

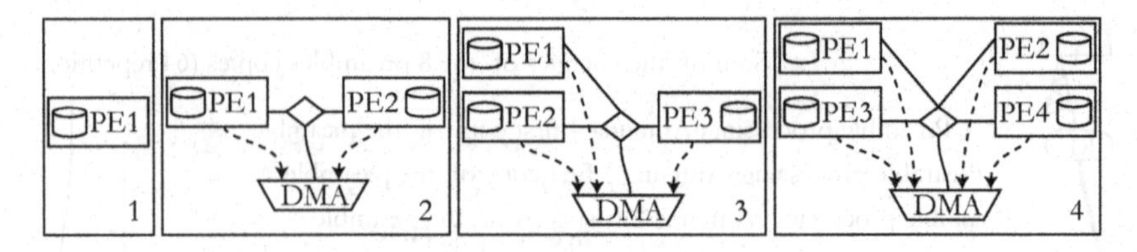

Fig. 7.4 Four architectures explored

representation in Fig. 7.3. After a total flattening (Sect. 3.5.2), the graph contains 2,209 vertices. This high number may be reduced by merging actors and/or reducing the flattening degree. Placing these actors onto the different cores by hand would be greatly time-consuming. As seen in Sect. 7.1.4 the rapid prototyping PREESM tool provides automatic scheduling, avoiding the problem of manual placement.

A significant point to note is that RACH decoding has an absence of feedback edge and a total independence of preceding preamble detection iterations. In the IBSDF graph description of RACH-PD, the graph is actually acyclic and the SDF to DAG transformation has no effect on its topology. The graph also displays a high parallelism that can be extracted at compile-time. Indeed, its topology depends only on parameters that are fixed for the lifespan of an eNodeB. Thus, no assignment step is needed at run-time to optimally execute RACH-PD on a distributed architecture. The parallelism can be evaluated in terms of span, length and potential speedup using the quality assessment chart from Sect. 6.6, computed in PREESM. However, this chart does not take communication costs or heterogeneity into account. An architecture exploration is needed to perform a precise and complete hardware and software codesign.

7.3.1 Architecture Exploration

The four architectures explored are shown in Fig. 7.4. The cores are all homogeneous Texas Instrument c64x+ DSP running at 1 GHz [17]. The connections are made via DMA-driven routes. The first architecture is a single-core DSP such as the TMS320TCI6482. The second architecture is dual-core, with each core similar to that of the TMS320TCI6482. The third is a tri-core and is equivalent to the tci6488 (Sect. 5.1.1). Finally, the fourth architecture is a theoretical architecture for exploration only, as it is a quad-core. The exploration goal is to determine the number of cores required to run the random RACH-PD algorithm in a 115 km cell and how to best distribute the operations on the given cores.

To solve the deployment problem, each operation is assigned an experimental timing (in terms of CPU cycles) in the scenario. These timings are measured with deployments of the actors on a single C64x+. Since the C64x+ is a 32-bit fixed-point DSP core, the algorithms must be converted from floating-point to fixed-point prior to these deployments. The EDMA is modeled as a parallel node controlled by a DMA (Chap. 5) transferring data at a constant rate and with a given set-up time.

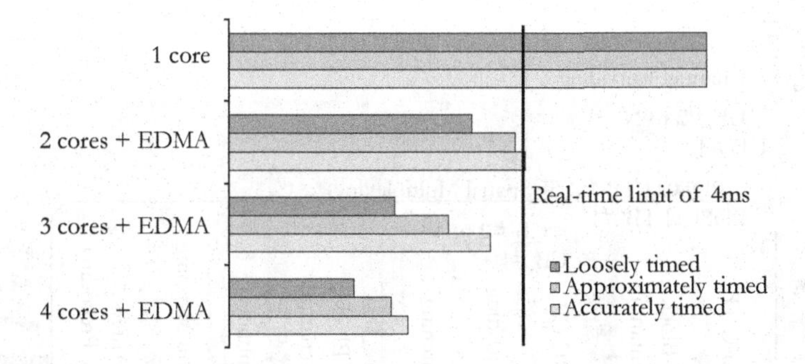

Fig. 7.5 Timings of the RACH-PD algorithm schedule on target architectures

The PREESM automatic scheduling process is applied for each architecture. The workflow used is close to that of Fig. 6.4. The simulation results obtained are shown in Fig. 7.5. The list scheduling heuristic is used with loosely-timed, approximately-timed and accurately-timed ABCs. Due to the 115 km cell constraints, preamble detection must be processed in less than 4 ms (Sect. 2.4.5).

The experimental timings were measured on code executions using a tci6488. The timings feeding the simulation are measured in loops, each calling a single function with L1 cache activated. For more details about C64x+ cache, see [17]. This warm cache case represents the application behavior when local data access is ideal and so will lead to an optimistic simulation. The RACH application is well suited for a parallel architecture, as the addition of one core reduces the latency dramatically. Two cores can process the algorithm within a time frame close to the real-time deadline with loosely and approximately timed models but high data transfer contention and high number of transfers make the dual-core architecture not sufficient when accurately timed model is used.

The 3-core solution is best: its CPU loads (less than 86 % with accurately-timed ABC) are satisfactory and do not justify the use of a fourth core, as can be seen in Fig. 7.5. The high data contention in this case study justifies the use of several ABC models; simple models are used for fast results and more complex models are then employed to correctly dimension the system.

7.4 Downlink Prototyping Model

Encoding the LTE downlink consists of preparing the different data and control channels and multiplexing them in the radio resources. Depending on the amount of control necessary in a subframe, the 1, 2 or 3 first symbols carry the control channels (Sect. 2.5.2). The remaining resources are shared between data channel (PDSCH), broadcast channels and reference signals . To encode the LTE downlink, the baseband processing retrieves the data Transport Blocks (TB) from the MAC layer (Fig. 2.6)

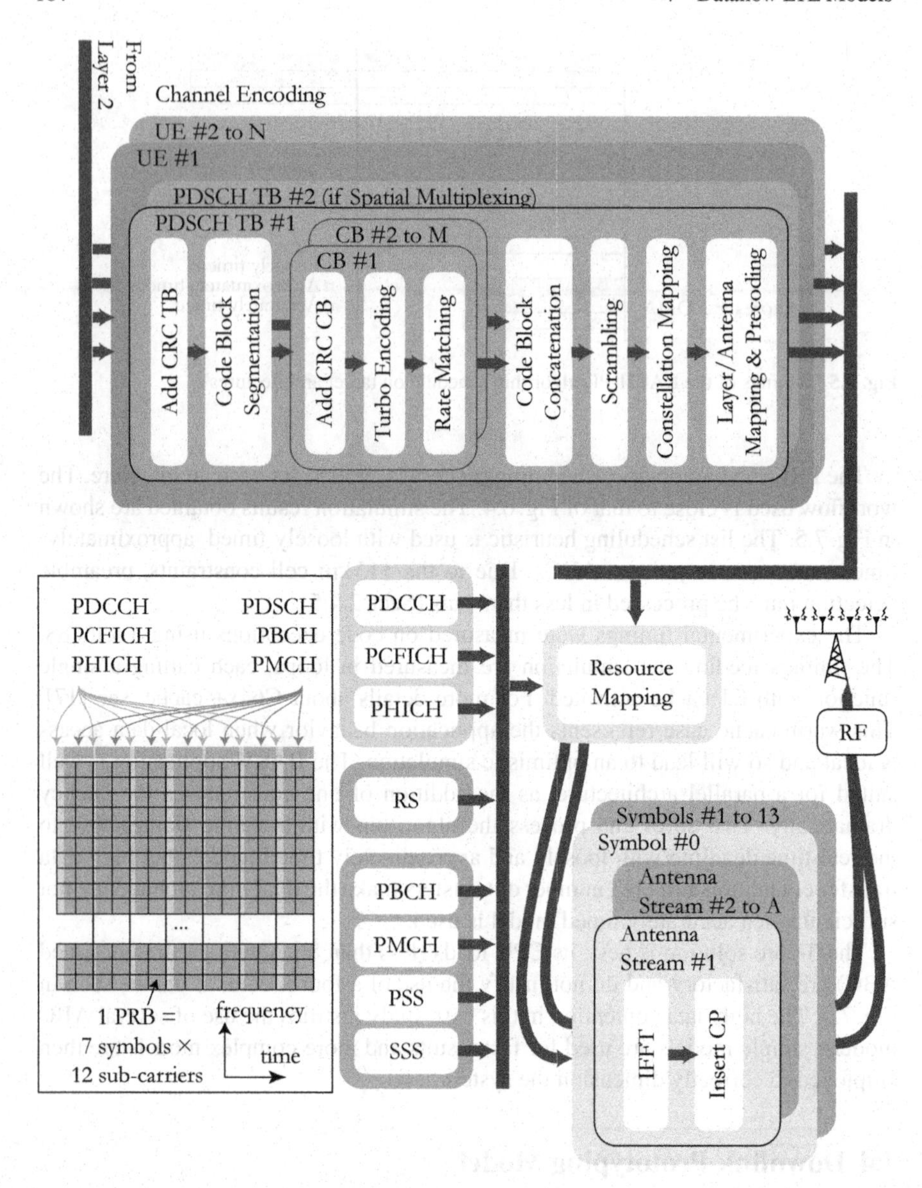

Fig. 7.6 Downlink decoding

and applies channel coding to them. Control channels and reference signals are then generated and multiplexed with the data channel prior to MIMO encoding.

A dataflow graph for encoding rapid prototyping is shown in Fig. 7.6. It is composed of three parts:

- During `channel encoding` the PDSCH data bits are transformed into symbols (Sect. 2.5.2). First, a Cyclic Redundancy Check (CRC) is added to each TB. The TB is then segmented into Code Blocks (CB) and each CB also receives a CRC. These CRCs enable the detection of errors when a UE decodes PDSCH. The resulting code block with CRC is Turbo-Encoded and rate-matched (Sect. 2.3.5) to introduce the exact amount of desired redundancy in the binary information for Forward Error Correction (FEC). Then, the encoded code-blocks are concatenated before being scrambled and constellation mapped. Scrambling spreads the redundant data evenly to protect it against the frequency-selective air channel and a constellation mapping is chosen among QPSK, 16-QAM and 64-QAM. The data is now composed of symbols. The decisions are not shown in Fig. 7.6 but all decisions (Hybrid ARQ parameters for retransmission, Modulation scheme, Antenna mapping mode,and so on) are sent by the MAC layer. Finally, the symbols are mapped to layers and the layers are precoded and mapped to antenna ports (Sect. 2.5.4). Layer mapping and precoding depend on the multiple antenna PDSCH mode (spatial multiplexing or not, open or closed loop) chosen by the MAC layer.
- During the `control and broadcast channels generation`, channel symbols are generated to be multiplexed with PDSCH. The primary control channel, PDCCH, as well as PCFICH and PHICH channels are mapped to the first 1, 2 or 3 symbols of the encoded subframe (Sect. 2.5.2). Certain Reference Symbols (RS) also appear in these first symbols. PBCH broadcast channel values are generated only if the current subframe contains cell information broadcast, which is the case once every frame of 10 ms. PBCH values are multiplexed with PDSCH, multicast channel PMCH, RS and Synchronization Symbols (PSS and SSS, Sect. 2.5.5) in the remaining 13, 12 or 11 symbols of the subframe.
- In the `front-end`, each complex value of the previously mentioned physical channels is associated with a Resource Element (RE) in the Resource Mapping procedure. Resource mapping is a bottleneck in the application because all symbols from all UEs must be collected prior to forwarding them to the rest of the front-end. An IFFT is then applied to convert each symbol of each antenna port into the time domain for transmission and a cyclic prefix is inserted between each symbol to reduce inter-symbol interferences (Sect. 2.3.2).

The PDSCH processing dominates the computation cost of other downlink physical channels. It is for this reason that it is divided into actors while PHICH generation is seen as a single “black box”. When described in an IBSDF, the number of repetitions of each actor (per TB, per CB, per UE, per antenna port…) is determined by the number of data tokens flowing between the different actors in the graph. Like the RACH-PD IBSDF description, the downlink encoding graph is fairly parallel and is already an acyclic graph. Unlike the RACH-PD IBSDF description, its topology depends on certain parameters (number of receiving UEs, number of TBs, size of these TBs and number of CBs) that vary each subframe because the subframe is the basic time unit for UE allocation (Sect. 2.3.2). It must be noted that only typical or worst-case downlink test cases can be studied in PREESM.

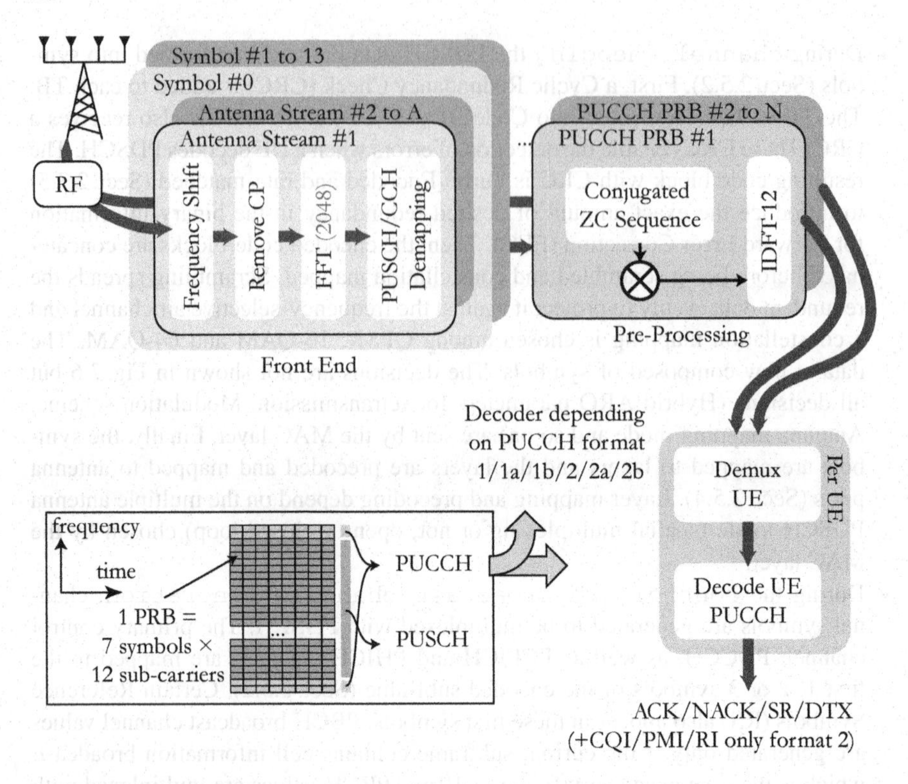

Fig. 7.7 PUCCH decoding

7.5 Uplink Prototyping Model

The uplink model is divided into PUCCH and PUSCH models. Decoding the PUCCH uplink control channel is a complex and potentially parallel operation. Contrary to the PDCCH encoding that may be represented as a single actor, as in Fig. 7.6, the PUCCH model is presented as a complex graph. It may be noted that the PUCCH and PUSCH graphs process the same raw symbols received by the antennas.

7.5.1 PUCCH Decoding

Decoding PUCCH consists of extracting resource blocks at the edge of the bandwidth from the raw data received by the antenna and then processing them to retrieve the uplink control bits. The localization of PUCCH in the bandwidth was explained in Sect. 2.4.2.

A dataflow graph for decoding PUCCH is displayed in Fig. 7.7. The PUCCH decoding consists in three distinct steps:

- The `front end` is common to PUCCH and PUSCH. The uplink subcarriers are frequency shifted by half the width of a subcarrier, i.e. by 7.5 kHz, to reduce distortions on the central resource blocks from the local oscillator. A frequency shift of 7.5 kHz is thus first applied to the received signal to compensate for this shift. The problem does not appear in downlink because one subcarrier (named d.c. subcarrier) is punctured (Fig. 2.21). In uplink, the same solution would worsen the PAPR and thus increase UE power consumption [12] p. 353. The CP is then removed and the symbols are transformed to the frequency domain by a FFT. The FFT size depends on the bandwidth (Table 2.1) with a maximum of 2,048 points in the 20 MHz bandwidth case with 1,200 subcarriers. Finally, resource elements from the band edges are used in the PUCCH decoding procedure while the center elements are used in the PUSCH decoding.
- The `pre-processing` step demultiplexes the PUCCH information from multiple UEs which was transmitted over the same resources using CDM (Sect. 2.4.1). For the UE, CDM encoding consists of multiplying, in the frequency domain, the data with a Zadoff-Chu Sequence of length 12 and with a given cyclic shift. Thus, it is possible for two UEs using the same ZC sequence with different cyclic shift to use the same PBR. For each PRB, there are twelve possible cyclic shifts. To decode the CDM, a multiplication with the Zadoff-Chu sequence is applied followed by a DFT of the size of a PRB.
- In the `decoder` step, the UE data is then demultiplexed and, depending on the PUCCH block format, certain information is extracted, using Reed-Muller decoding to decode CQI (Sect. 2.3.5). ACK is reported by the UE in the case of a correct downlink reception; NACK for a failed downlink reception (when the CRC indicated errors in transmission) and DTX means that the UE did not even detect in PDCCH that it had data to receive in PDSCH.

The information extracted from PDCCH is reported to the MAC layer, which controls the downlink communication. The PDCCH can also contain an uplink Scheduling Request (SR) when a UE requires more PUSCH resources.

7.5.2 PUSCH decoding

PUSCH decoding is the most complex graph in the LTE eNodeB physical layer model. It contains seven parts and is illustrated in Fig. 7.8:

- The `front end` is common to PUCCH and PUSCH and was explained in Sect. 7.5.1. It outputs the physical resource blocks dedicated to PUSCH in frequency domain.
- The `data/DM RS/SRS demultiplexer` separates the reference signals from the data resource elements.
- The `channel sounding` processes the SRS symbol 13 and sends sounding information to the MAC layer (Sect. 2.4.1). This information is used to schedule UEs in the subsequent subframe.

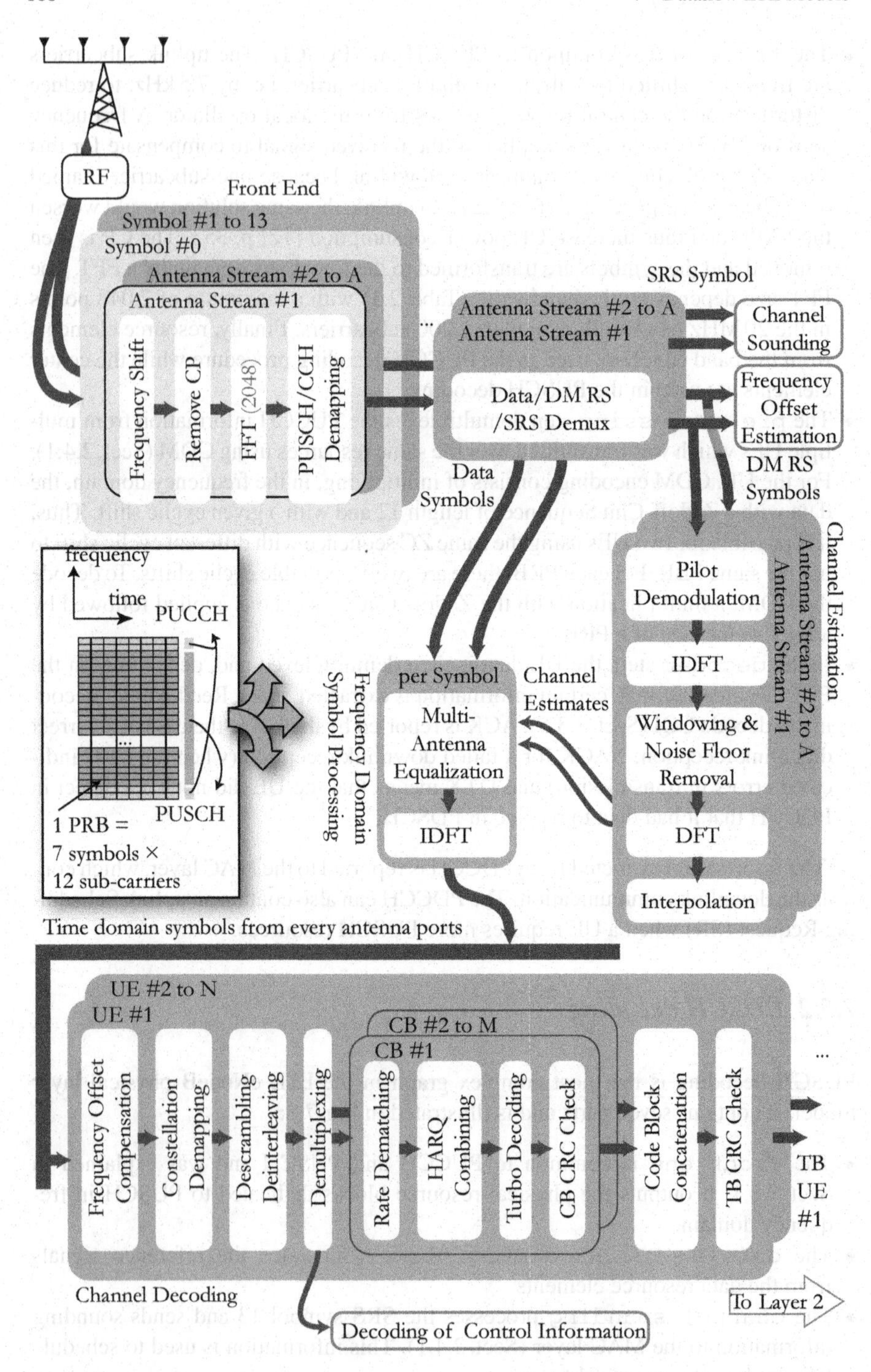

Fig. 7.8 PUSCH decoding

- The `frequency offset estimation` evaluates the Doppler effect experienced by each UE data TB due to UE velocity. The estimation uses the two reference DM RS symbols 3 and 10 (Sect. 2.3.1). The frequency offset is then transmitted to the MAC layer and influences the channel decoding.
- The `channel estimation` processes the ZC sequences from the two DM RS symbols to compute the Channel Impulse Response (CIR). The pilot subcarriers from different UEs are first separated. They are then transformed into time domain by an IDFT. The DM RS signal is synchronized and noise removed. Finally, the signal is transformed back into frequency domain, the CIR is evaluated and is linearly interpolated to find the CIR of each of the eleven data symbols.
- The `frequency domain symbol processing` first combines the data from different antenna ports, equalizing them (Sect. 2.3.2) and using either MRC or MIMO decoding, depending on the UE multiple antenna scheme employed (Sect. 2.3.6). As with channel estimation, the data from each UE is processed independently because it may have different MCS that other UEs. The data is then transformed into time domain by an IDFT. It may be noted that the equalizer processes the data from all antenna ports together. Moreover, after Frequency Domain Symbol Processing, the data from all the symbols must be gathered for joint processing. These two bottlenecks limit functional parallelism, introducing additional causality in the system but they cannot be avoided.
- The `channel decoding` step is approximately the inverse operation of the downlink channel encoding explained in Sect. 7.4. The transport block values of each UE are processed, extracting bits from the time-domain symbols. The frequency offset due to the Doppler effect is compensated and data is demapped, descrambled and deinterleaved. A second demultiplexing of data and control is necessary at this point because certain control values can be multiplexed with data if PDCCH capacity is insufficient [12] p. 398. Each CB is then dematched and combined with any previous HARQ repetition. A hidden feedback edge exists because code blocks from previous iterations of the graph are maintained in the MAC layer for future combination with the subsequent repetition. After Turbo decoding, the CRC is checked to verify whether data was lost and a new HARQ repetition is necessary. CBs are finally gathered and the TB CRC is checked.

Contrary to the downlink part, only one TB per UE can be received in a subframe because no spatial multiplexing other than MU-MIMO is allowed. In order to obtain a correct simulated decoding latency, the IBSDF graph must account for the fact that each symbol is received with a delay of 1/14 ms more than the preceding one. As there is no notion of time in IBSDF and it is the scenario alone that introduces timings for actors, a "trick" must be used to include these delays. This "trick" consists of introducing a fake processing element in the architecture, which this study will call Antenna Interface (AIF). A "dummy" delay actor is added to the algorithms, assigned to AIF, associated with a 1/14 ms timing on AIF in the scenario. This actor is preceded by another "dummy" actor whose use is to repeat the delay. A feedback edge on the delay actor ensures the order of execution of its instances. The original graph and its form after transformation to a single rate graph are shown in Fig. 7.9.

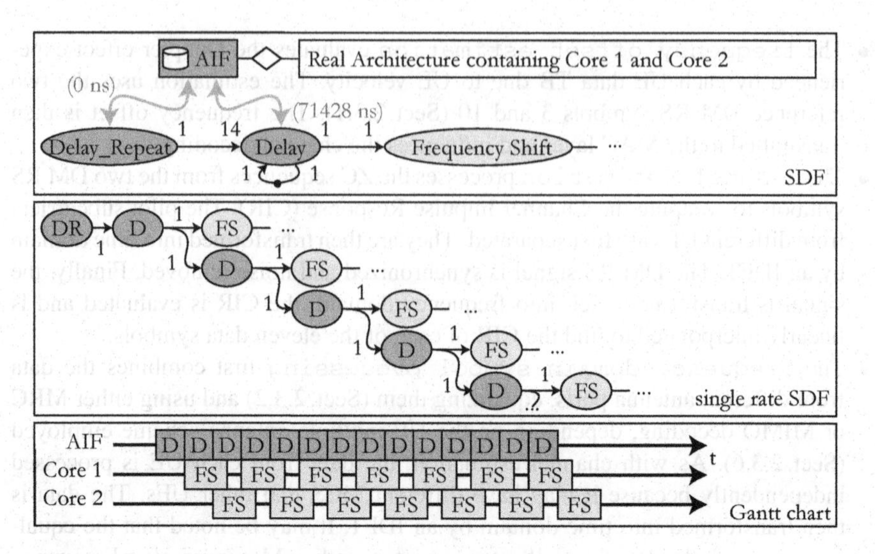

Fig. 7.9 Introducing delays in a IBSDF description to simulate events

The delay from the first "dummy" actor must be subtracted from the latency. The same technique can be used in the downlink model to receive transport blocks at disparate times.

It may be noted that the data rate of both uplink and downlink communications between baseband processing and antenna ports can be a challenge. For a 20 MHz cell with 30.72 MHz sampling rate and 4 antenna ports and with complex data of 16-bit real and imaginary parts, the antenna data rate is almost 4Gbit/s in both directions. Consequently, the hardware antenna interface needs to support very high data rates. The complex AIF data exchanges can be studied at system-level using PREESM and S-LAM Contention Nodes, without considering all the details of implementation.

Of course, when the RACH-PD, downlink and uplink graphs are all prototyped together, the result is an extensive graph. Synchronization can then be specified for the reception time of the downlink transport blocks and also the reception time of the uplink symbols, using the "trick" presented above. This chapter presented LTE modelization for rapid prototyping. The next chapter explains code generation from LTE dataflow graph descriptions.

References

1. Pelcat M, Piat J, Wipliez M, Nezan JF, Aridhi S (2009) An open framework for rapid prototyping of signal processing applications. EURASIP J Embed Syst
2. Eclipse Open Source IDE: http://www.eclipse.org/downloads/
3. Gamma E, Helm R, Johnson R, Vlissides J (1995) Design patterns: elements of reusable object-oriented software. Addison-wesley Reading, Boston

4. SDF4J: http://sourceforge.net/projects/sdf4j/
5. Sarkar V (1987) Partitioning and scheduling parallel programs for execution on multiprocessors. Ph.D. Thesis, Stanford University
6. Brandes U, Eiglsperger M, Herman I, Himsolt M, Marshall MS (2001) Graphml progress report, structural layer proposal. In: Mutzel P, Junger M, Leipert S (eds) Graph Drawing-9th International Symposium, GD Vienna, 2001. Springer, Heidelberg, pp 501–512
7. Eker J, Janneck JW (2003) CAL Language Report. Technical Report, ERL Technical Memo UCB/ERL M03/48, University of California at Berkeley
8. Janneck JW (2007) NL - a Network Language. Technical Report. ASTG Technical Memo, Programmable Solutions Group, Xilinx Inc.
9. w3c XSLT standard. http://www.w3.org/Style/XSL/
10. Grammatica parser generator: http://grammatica.percederberg.net/
11. Janneck JW, Esser R (2001) A predicate-based approach to defining visual language syntax. In: Symposium on visual languages and formal methods, HCC01, Stresa, pp 40–47
12. Sesia S, Toufik I, Baker M (2009) LTE the UMTS long term evolution from theory to practice. Wiley, New York
13. 36.213 G.T.: Evolved Universal Terrestrial Radio Access (E-UTRA); physical layer procedures (Release 9) (2009).
14. Mehlführer C, Wrulich M, Ikuno JC, Bosanska D, Rupp M (2009) Simulating the long term evolution physical layer. In: Proceedings of the 17th european signal processing conference (EUSIPCO 2009). Glasgow, (2009)
15. Jiang J, Muharemovic T, Bertrand P, Random access preamble detection for long term evolution wireless networks. Patent 20090040918
16. 36.211 G.T.: Evolved Universal Terrestrial Radio Access (E-UTRA); physical channels and modulation (Release 9) (2009)
17. TMS320C64x/C64x+ DSP (2008) CPU and instruction set reference guide, texas instrument technical document (SPRU732G)

Chapter 8
Generating Code from LTE Models

8.1 Introduction

Literature on automatic multi-core code generation was reviewed in Sect. 4.5 and scheduling strategies in Sect. 4.4.1. In this section, generated code execution schemes are defined, detailing how code is generated from a given scheduling strategy. Code generation for RACH-PD (Sect. 8.2), PUSCH (Sect. 8.3) and PDSCH (Sect. 8.4) is then explained. Combining the three dataflow paths is discussed finally in Sect. 8.5.

8.1.1 Execution Schemes

In [1], Bell and Wood advise programmers on multi-core programming. The architectures they study are the same DSPs targeted by this study. Their recommended method for parallelizing applications is based on a test-and-refine approach and is manual with steps that are different from those of Fig. 4.2. The report then presents the two execution schemes which form the basis of distributed code generation: the master/slave scheme and the dataflow scheme. Instead of "dataflow" scheme, the terminology "decentralized scheme" will be used to avoid confusion with dataflow MoC. An execution scheme differs from a MoC in that it defines the architecture of a running code, with run-time system considerations that MoCs do not model.

- The `master/slave scheme`, illustrated in Fig. 8.1a, consists of a centralized execution control code in a master task that posts actors to slaves tasks. In this model, actors are often called jobs but this term is usually used for actors which are loosely coupled with respect to each other. A master or a slave task can be entirely software-defined or have certain parts that have been hardware accelerated. Each task can be locally contained in threads controlled by an OS or be the only computation of an operator. The centralized control code simplifies the efficient distributed execution of the various applications, i.e applications are dominated by the control code rather than data because these applications require a

M. Pelcat et al., *Physical Layer Multi-Core Prototyping*,
Lecture Notes in Electrical Engineering 171, DOI: 10.1007/978-1-4471-4210-2_8,

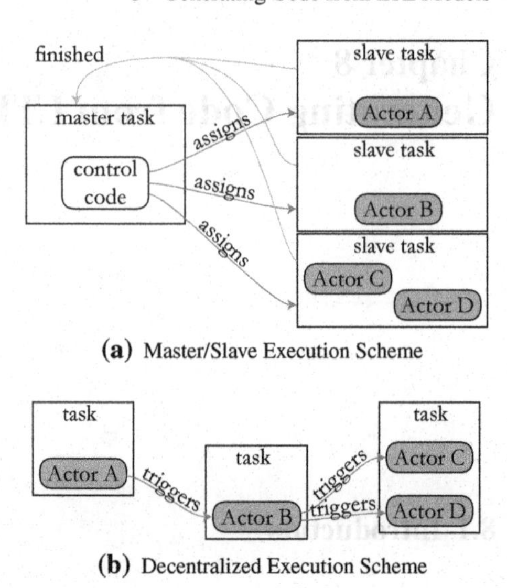

Fig. 8.1 Execution schemes

(a) Master/Slave Execution Scheme

(b) Decentralized Execution Scheme

centralized knowledge of the control values to efficiently reconfigure the parallel execution. [1] advises use of the master/slave scheme for RLC and MAC Layers of telecommunication systems (Sect. 2.2.2), as these layers are always dominated by the control code. For instance, in LTE, the eNodeB MAC scheduler is a complex control-dominated operation which varies greatly depending on the state of the communication (number and distance of the connected users, quality of the channels, MIMO schemes, and so on). A master can either be an "intelligent" unit, posting actors where they will be most efficiently executed or a simple pool of actors that the slaves monitor for actor availability.

- The `decentralized scheme`, illustrated in Fig. 8.1b, consists of independent tasks which wait for input data, process it and send it without use of a global execution arbitrator. This scheme corresponds to a "real" dataflow implementation where computation is triggered solely by data arrival. Naturally, this scheme is well suited to dataflow application implementation and [1] advises the use of the decentralized scheme when implementing the physical layer of telecommunication systems. However, as the uplink and downlink streams of the LTE eNodeB physical layer are quite balanced between data and control, the most appropriate execution scheme for these algorithms is clearly a mix of the master/slave and decentralized schemes.

As centralized resources in distributed architectures usually form system bottlenecks, the centralization of the master in the master/slave scheme is a real issue. `Hierarchical approaches` can reduce the problem. In a hierarchical multicore scheduler, a top-level scheduler assigns a pool of operators to a part of the application. The assignment and scheduling of that subsystem is then managed by a lower-level scheduler which is located elsewhere. Pools of operators can overlap;

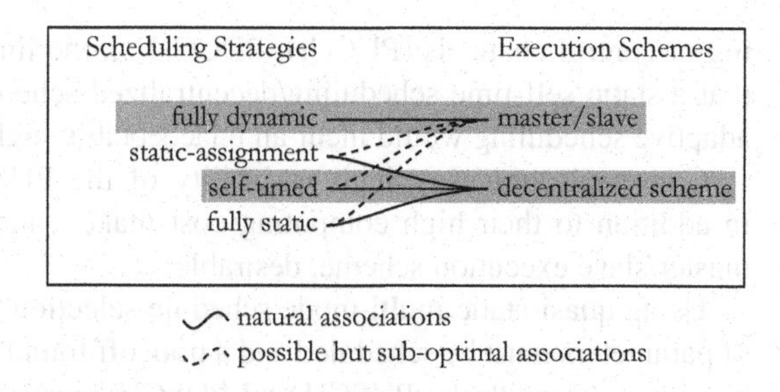

Fig. 8.2 Possible associations of scheduling strategies to execution schemes

in this case, a time-sharing scheme is necessary for each shared operator. This is usually performed by the top-level scheduler reconsidering its choices periodically. System introspection (monitoring operator activity) can also be used to choose when to reconsider top-level assignment. [2] is an example of such a hierarchical scheduler with two levels.

Both master/slave and decentralized schemes can be used to generate code from dataflow MoCs. However, the most appropriate execution scheme depends on the chosen scheduling strategy (Sect. 4.4.1). Static-assignment, self-timed and fully static scheduling strategies can be implemented using both master/slave and decentralized schemes. However, all three strategies execute the assignment step at compile time to simplify run-time and avoid the need for a centralized control code. Consequently, the most efficient execution scheme for static-assignment, self-timed and fully static scheduling strategies is the decentralized scheme. Conversely, the fully dynamic scheduling strategy necessitates a run-time actor assignment. This strategy demands the centralization of either the execution control code or the pool of actors.Therefore, the master/slave scheme is the only possible execution scheme for this scheduling strategy. The relation between scheduling strategies and execution schemes is illustrated in Fig. 8.2 and the two pairs of strategy/scheme that this study uses for LTE are highlighted.

A master element containing a multi-core scheduler is the central piece of a multi-core RTOS. Such a system, if highly optimized, could improve embedded multi-core programming by offering an abstraction of heterogeneous computing not available today.

8.1.2 Managing LTE Specificities

For each dataflow graph, the right scheduling strategy needs to be selected depending on the variable behavior of target application and architecture. Once the scheduling strategy is chosen, the execution scheme is chosen, as seen in Fig. 8.2. For the RACH-PD algorithm, the couple self-time scheduling/decentralized scheme is the most suitable solution due to static system behavior (Sect. 7.2) as well as the fact that the execution time of the target operators cannot be precisely known. For the processing of control channels (PUCCH, PDCCH...), the limited costs of the actors mean that a static self-time scheduling/decentralized scheme is a good solution because adaptive scheduling would incur an unacceptably high overhead.

Conversely, the very high variability of the PUSCH and PDSCH algorithms in addition to their high computing cost make a more complex strategy, such as master/slave execution scheme, desirable.

Using quasi-static multi-mode schedule selection [3], all possible combinations of parameters must be scheduled and a tradeoff found at compile-time, which would be impossible with the PUSCH and PDSCH because of the high number of cases (Sect. 8.3.4.1). PUSCH and PDSCH necessitate the use of run-time adaptive scheduling instead of multi-mode compile-time scheduling. These two code generation ideas are developed in the following sections.

8.2 Static Code Generation for the RACH-PD Algorithm

The RACH-PD is a static algorithm, so it can be efficiently parallelized using the PREESM framework code generation.

8.2.1 Static Code Generation in the PREESM Tool

The PREESM tool code generation uses the `self-timed scheduling strategy` (Sect. 4.4.1) and the `decentralized execution scheme`. Many principles are adapted from the AAM methodology and incorporated into the tool [4]. The code generated by the PREESM tool is analogous to a coordination code (Sect. 3.2). The host code and communication libraries (Sect. 3.2) must be manually coded.

To generate code statically, the static scheduling of the input algorithm on the input architecture must have been defined. A typical PREESM workflow performing code generation is displayed in Fig. 6.4. For each operator, the generated code consists of an initialization phase and a loop endlessly repeating the part of the algorithm graph assigned to this operator. Special code loops are generated from IBSDF descriptions for the non-flattened and repeated hierarchical actors in order to obtain a compact code [5]. Each atomic actor generates a function call on its assigned core as explained in Sect. 6.2.1. The function call prototype is defined in an IDL file. The path of the IDL files is a parameter of the code generation workflow node. From the deployment information provided by the scheduler, the code generation module creates a generic representation of the code in XML. This code is then XSLT-transformed to produce the specific code for the target. The code generation flow for a tri-core tci6488 target processor (Sect. 5.1.1) is illustrated in Fig. 8.3.

The XML generic code representation is illustrated by an example in Fig. 8.4, which shows the XML code file generated for the operator op1 of Fig. 8.3. The file

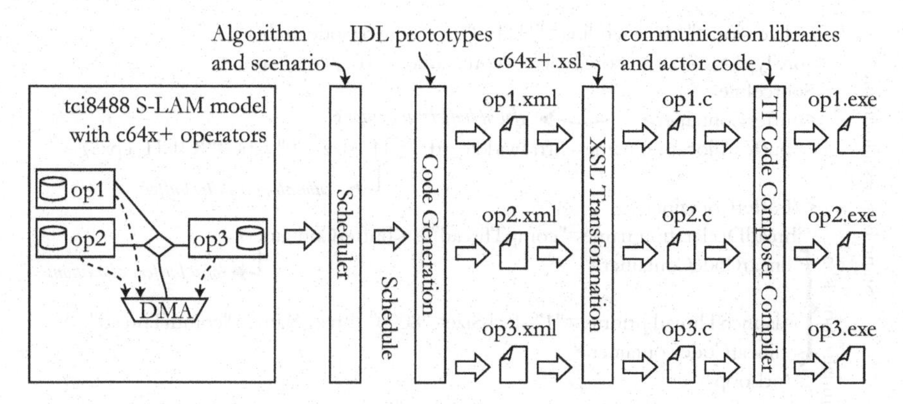

Fig. 8.3 Code generation procedure on a tci6488

contains a static buffer allocator, a computation thread and a communication thread. It also contains an operator type which indicates which XSLT transformation file to use when generating the coordination code, in this case "c64x+.xsl". The two threads communicate via statically allocated shared buffers and semaphores. The communication thread controls the DMA and waits for completion of transfers, while the computation thread executes code in parallel, calling functions with parameters of type constant or buffer.

PREESM currently supports the C64x and C64x+ based processors from Texas Instruments with a DSP-BIOS Operating System [6] and the x86 processors with a Windows Operating System. Device-specific code libraries have been developed to manage the communications and synchronizations between the target cores [7]. The XSLT transformation generates calls to the appropriate predefined communication library, which are dependent on the route type and on the names of the communication nodes between operators in the S-LAM architecture model (Chap. 5). The inter-core communication schemes supported include TCP/IP with sockets, Texas Instruments EDMA3 [7] and RapidIO link [8].

Figures 8.5 and 8.6 show the behavior of a very simple application with three actors A, B and C when mapped on two cores of a tci6488. In Fig. 8.5a, the actors are assigned to operators. In Fig. 8.5b, a graph of the code execution is shown where the data and the semaphores are displayed. The data is sent via a shared buffer that, combined with semaphores, implements a FIFO queue of size 1. Semaphores need to synchronize the transfer, successively informing the receive operator that the buffer is full and the sender that the buffer is empty and ready for the next iteration. The initial token on the feedback semaphore edge in Fig. 8.5b shows that the semaphore signaling an empty buffer targets the next iteration of the graph. Also in Fig. 8.5b, the send and receive operations are gathered in one actor because they are synchronized by the communication library. The general problem of bounding FIFOs for executing SDF graphs is discussed in Chap. 10 of [9]. The solution employed is to add feedback edges with tokens to limit the number of firings of the actors that are otherwise always

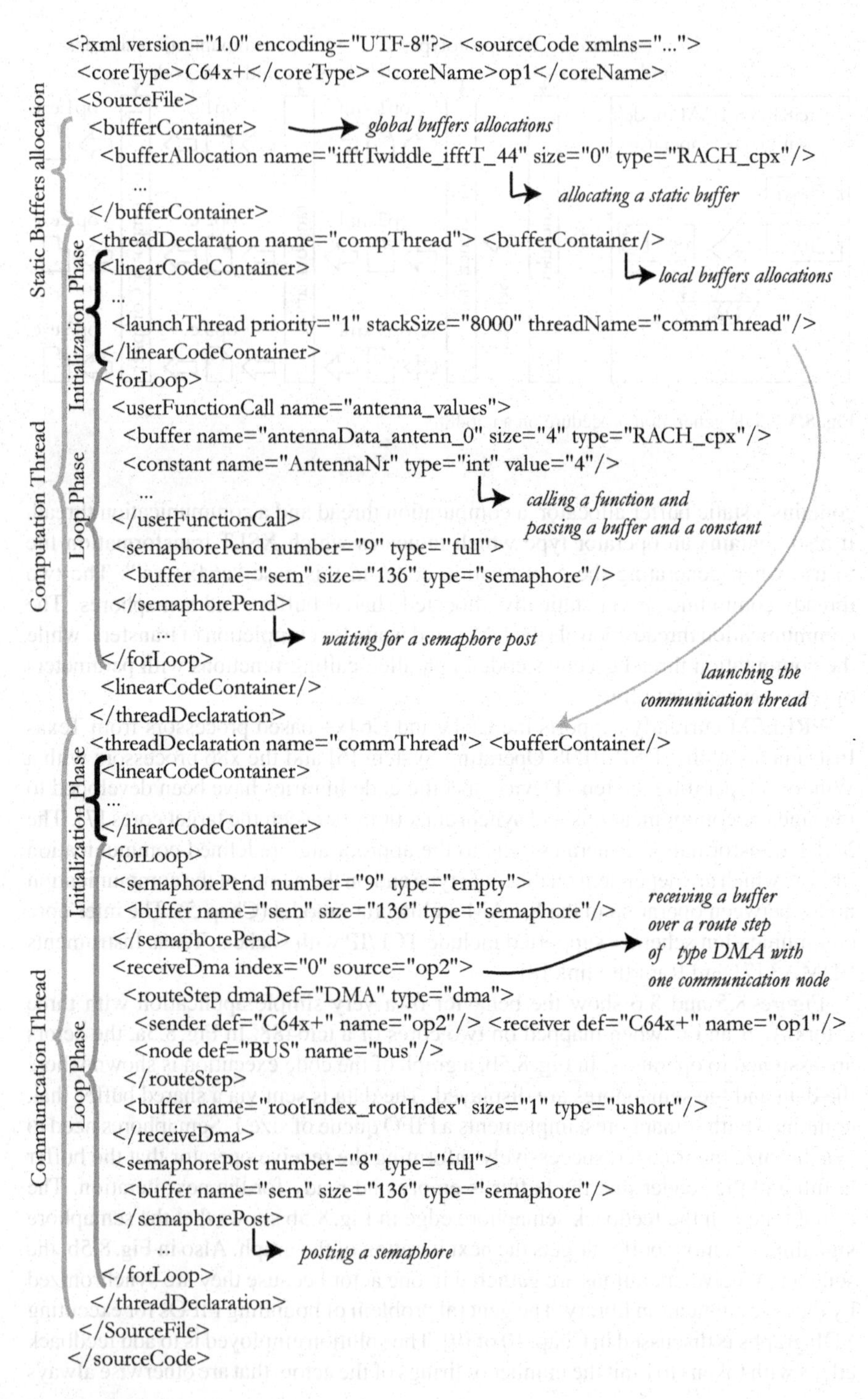

Fig. 8.4 XML generic code representation

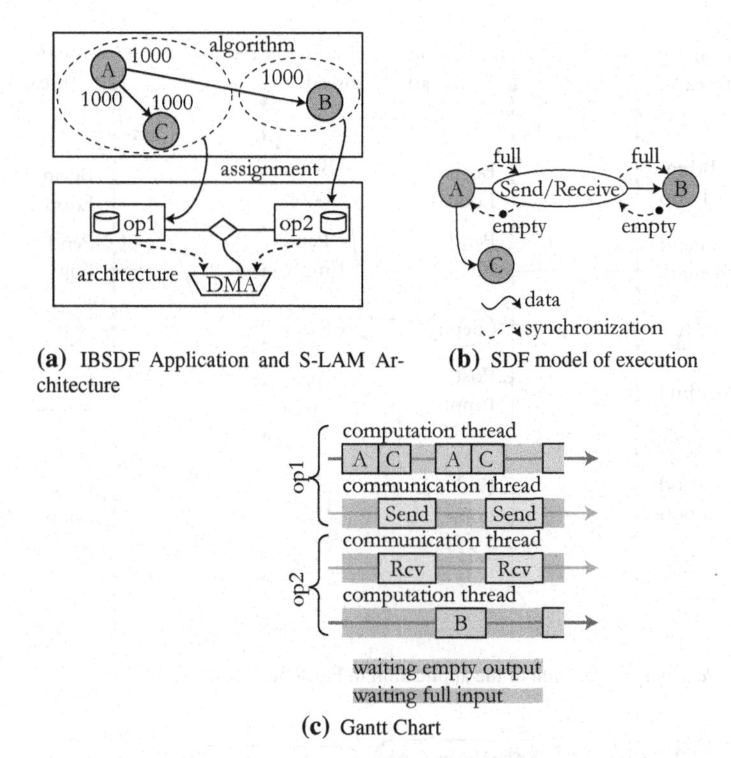

(a) IBSDF Application and S-LAM Architecture

(b) SDF model of execution

(c) Gantt Chart

Fig. 8.5 Code Behavior of a an example of message passing with DMA

free to fire. For example, an actor with no input edge could fire many times before any other actor fires, piling many tokens in its output FIFO queues.

Figure 8.5c shows a Gantt chart of the graph execution where *C* can be executed in parallel with a transfer from operator 1 to operator 2 because of the synchronization between the DMA and the communication thread. The semaphores synchronizing communication and computation threads need to be initialized during code initialization phase. Figure 8.6 shows, as a Petri net (Sect. 3.1.1), the executions of the two threads running on each core of the architecture. The petri net details clearly the accesses to semaphores; this representation is often used to represent code generation synchronization in the Algorithm Architecture Matching (AAM) method [10]. Figures 8.6 and 8.5b show equivalent representations, one typical from the AAM method and the other using dataflow graph graphic conventions.

8.2.2 Method Employed for the RACH-PD Implementation

The development of the communication library to transfer data using EDMA3 (Sect. 5.1.1) on a tci6488 is explained in [7]. The complete code generation of

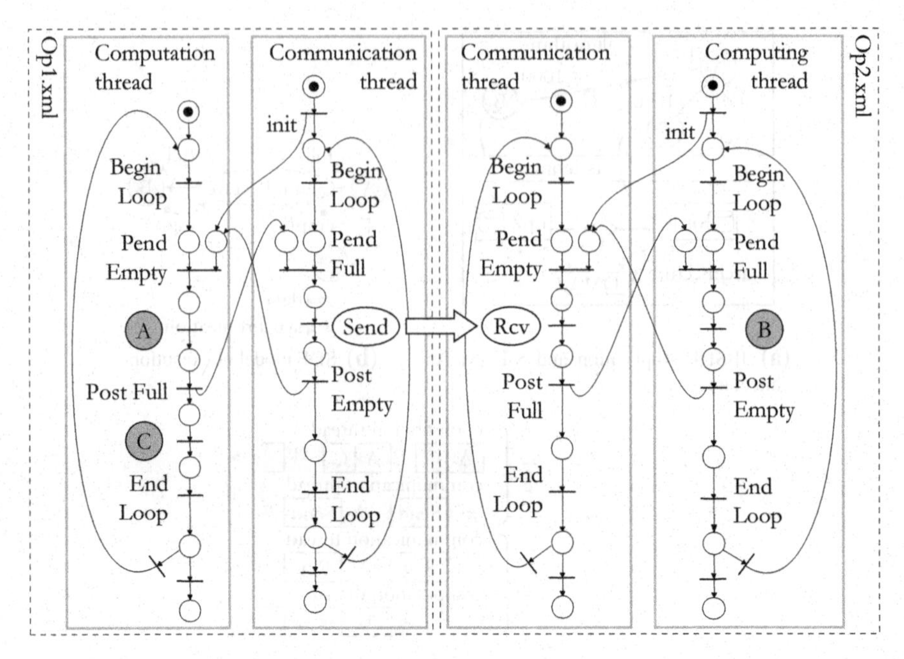

Fig. 8.6 Petri net of execution of the application in Fig. 8.5a

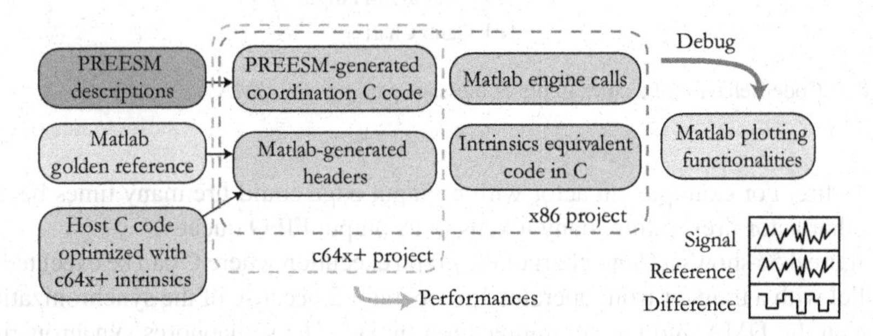

Fig. 8.7 Method for the RACH-PD algorithm implementation

RACH-PD is also detailed. From a Matlab model of the RACH encoding and decoding, a C host code of the RACH-PD and the generated coordination code, a highly efficient multi-core development and debugging method can be constructed. The debugging phase can be fully executed on a single-core PC, using Matlab to generate test vectors and display the results, and using PREESM to generate coordination code. The efficient implementation of the embedded multi-core deployment is then generated seamlessly. This method, illustrated in Fig. 8.7, was successfully applied to RACH-PD. It can be divided into six steps:

1. RACH-PD functions are written in C, inspired by a Matlab "golden" code which encodes and decodes the PRACH channel. Each C function of the RACH-PD

embedded code is hand-written, adding c64x+ intrinsics to optimize for speed. Intrinsics specify assembly calls in a C code. Each function is then instrumented and run independent of target to collect execution timestamps and allow evaluation of their cost.
2. PREESM generates two coordination codes from the IBSDF description of RACH-PD illustrated in Fig. 7.3 and two S-LAM architecture models: a single-core x86 and the targeted multi-core embedded platform. The x86 architecture will serve for debugging phase and the target for the implementation. The actor timings previously measured are used to simulate the execution while scheduling the graph.
3. Matlab encodes an input signal containing RACH preambles and then decodes it. It generates large header files in C code containing buffers with the data at several steps in the decoding process, which are displayed in Fig. 7.3, such as before and after FIR, before and after correlation, before and after IFFT, and so on.
4. The code is run on a x86, adding C reference code for each c64x+ intrinsic and linking to the Matlab engine for displaying the results in Matlab. The decoded signal is contained in a Matlab-generated header. At each decoding step, a separate header is generated by Matlab, and then the golden reference and the embedded solution are plotted using Matlab engine. Code errors and insufficient data accuracy are clearly visible in the plots, which can thus be used to debug the code.
5. Communication libraries are hand-written and then debugged for the target (alternatively, appropriate communication libraries may be retrieved if a previous project has used the same architecture).
6. Maintaining the identical host code but switching the coordination code, the RACH-PD is run on the target architecture to test its speed and accuracy of its results. Both the intrinsic equivalent C code and the Matlab engine calls are then removed from the embedded project. No deadlock will appear at this point, as data dependencies are handled during only automatic scheduling.

The above method has several benefits. Firstly, the code can be debugged without consideration of parallel programming problems. Secondly, the debugging phase is faster than debugging with the target board because it involves no cross-compilation. Lastly, the parallelism has been entered into the PREESM framework which facilitates the redeployment of the algorithm on new architectures. These advantages make static code generation a powerful tool for creating embedded multi-core code.

Code has been generated and tested for the RACH-PD algorithm. The resulting implementation is deadlock-free but needs many manual tweaks to reduce its synchronization and memory consumption and the final result is compatible with the simulated results from Fig. 7.5 in terms of cadence (a cadence under 4 ms is obtained) but not in terms of latency. Some memory and synchronization optimizations are still needed in PREESM to obtain automatically an optimized code.

The static code generation process is efficient only if the actor times are relatively stable and the coordination topology is absolutely stable. This is the case for RACH-PD but not for PUSCH decoding and PDSCH encoding. In the next section, a technique to automatically schedule PUSCH on multi-core embedded systems is developed.

8.3 Adaptive Scheduling of the PUSCH

The underlying goal of adaptive scheduling is to create the central part of an efficient multi-core RTOS for signal processing applications on heterogeneous architectures. An adaptive scheduler needs to schedule its application efficiently with very little overhead. The implemented system is dedicated to a specific task. Consequently, the scheduler can adapt at compile time or during system launch to the system specificities and run efficiently afterwards. The adaptive scheduler implements a `fully dynamic scheduling method` and is an element of the master entity in a `master/slave execution scheme`. A difference of the adaptive scheduler when compared to usual operating systems is that the adaptive scheduler manipulates actors instead of threads and should not need preemption, if actors are small enough, to respect application real-time constraints. Context switching is thus useless for adaptive scheduling and deadlocks or race conditions do not exist.

In Sect. 4.4, adaptive scheduling approach was introduced. Adaptive scheduling is a fully dynamic scheduling method which assigns actors to operators at run-time and orders them using simple heuristics. Processing the LTE PUSCH in the eNodeB consists of receiving the multiplexed data from connected UEs, decoding it and transmitting it to the upper layers of the standard stack (Sect. 7.5). Depending on the number of active UEs and on their instantaneous data rates, the decoding load may vary dramatically. A multi-core adaptive scheduler has capacity to recompute a schedule for the algorithm every millisecond. One important function of adaptive scheduler is to maintain low uplink latency. The LTE standard has strict constraints in terms of latency, limiting the available time for both uplink and downlink processing. The scheduler uses a graph modeling technique, as during the rapid prototyping phase, to determine the execution timing from the DAET of each actor on each operator. The schedule is determined after an "on-the-fly" simulation of the application execution. The next section introduces the PUSCH model for run-time scheduling.

8.3.1 Static and Dynamic Parts of LTE PUSCH Decoding

The PUSCH decoding operation (Fig. 7.8) can be divided in two parts :

1. The static part: `FFT front end processing and multiple antenna Minimum Mean Square Error (MMSE) equalization`. The parameters of these two operations (number of receive antennas, Frequency Division Duplex (FDD) or Time Division Duplex (TDD), bandwidth, and so on) are fixed during run-time. The Cyclic Prefix (CP) is removed, the frequency

is shifted by 7.5 kHz as stipulated by the 3GPP LTE standard, the symbols are converted into frequency domain by a Fast Fourier Transform (FFT) and equalized using received reference signals. The data from up to four antennas is then combined and the subcarriers are reordered. Finally, an Inverse Discrete Fourier Transform (IDFT) reconverts the data back into the time domain per user basis.
2. The dynamic part: `channel decoding`. For this operation, the parameters (number of connected UEs, number of allocated Resource Blocks, modulation order, and so on) are highly variable during run-time. The multi-core scheduling of this dynamic part must be adaptive.

The first and static part of processing can be represented as an IBSDF graph and demonstrates a high level of parallelism. Consequently, the corresponding self-timed parallel schedule can be processed at compile time using PREESM [11], which will also generate a self-timed code similar to that generated for RACH-PD (Sect. 8.2). The multi-core scheduling of the second and dynamic part needs to be adapted at run-time to the varying parameters. A discussion of the constraints on these varying parameters follows in the next Section.

8.3.2 Parameterized Descriptions of the PUSCH

The parameters of PUSCH decoding are specified in the eNodeB MAC layer. They are available at least 1 ms before the start of the decoding. This property makes parameterized dataflow (Sect. 3.5.1) particularly suitable for describing PUSCH decoding because this millisecond allows the construction of an execution graph and the subsequent search for an efficient multi-core schedule using this graph.

The most common use of parameterized dataflow is to describe an algorithm in a hierarchical PSDF graph where init and sub-init phases ϕ_i and ϕ_s modify topology and actor parameters at different levels. Such a PSDF description of the PUSCH decoding is displayed in Fig. 8.8. In this usage, the static part is not developed because it is parallelized at compile-time (Sect. 8.3.1). The dynamic part is simplified compared with the graph displayed in Fig. 7.8: actors have been merged to reduce the final size of the graph. Figure 8.8 is divided into two actors at top level: the `convergence` actor that gathers data from all subframe symbols and the `ChannelDecoding` actor. The top-level initialization phase ϕ_i, is executed once per graph invocation (i.e. once per millisecond), and initializes parameters carrying the maximum number of Code Blocks (CB) per UE, and the maximum size of the CBs. In the sub-init phase ϕ_s, the number of UEs is retrieved from the MAC layer before each graph execution. The channel decoding actor is repeated nb_UE times, as many times as there are UEs sending data in the subframe, as a consequence of token productions and consumptions. The channel decoding actor contains a hierarchy with five actors. The `keepCurrentTones` actor filters the subcarriers used for data transmission and transmits the resultant subcarriers to the `perUEProcessing` which processes the frequency offset compensation, channel

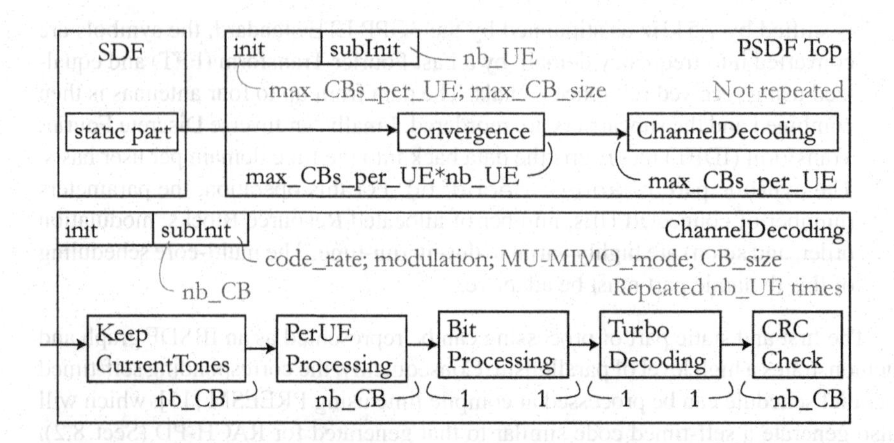

Fig. 8.8 PSDF description of the uplink

decoding, constellation demapping, descrambling, deinterleaving and demultiplexing. The `BitProcessing` actor gathers rate dematching and HARQ combining information while the `Turbo Decoding` actor associates turbo decoding with CB CRC checking. Finally, the `CRCCheck` actor checks the CRC of the TB. At each invocation of ChannelDecoding, which occurs for each UE, the number of CBs is modified as well as the size of these CBs. For a topology parameter to be modified before each iteration of the graph, it must be modified in the sub-init graph ϕ_s of ChannelDecoding to be executed before each execution of the body ϕ_b. Some actors parameters are also modified in ϕ_s, such as code rate, modulation, MU-MIMO mode.

In all the LTE graphs of the current study, there is no feedback path in the physical layer. The SDF graph in Fig. 8.8 is a Directed Acyclic Graph (DAG, Sect. 3.2.3). Knowing all UE and CB parameters in advance of a subframe allows the global reconfiguration of the PUSCH graph instead of reacting locally to the variations of parameters in sub-init graphs such as in Fig. 8.8. The benefit of such a global reconfiguration is to provide a system view of subframe decoding and to enable a simulation of the code execution on a heterogeneous multi-core architecture using the compile-time methods presented in Sect. 4.4. To achieve a global graph, patterns of token production and consumption are used as in Parameterized Cyclo Static Dataflow Graphs (PCSDF) Sect. 3.4. A run-time system cannot handle the Basis Repetition Vector (BRV) computation of a PCSDF graph to instantiate the actors. The input model of the adaptive scheduler is thus reduced to acyclic graphs and called Parameterized Cyclo Static Directed Acyclic Graph (PCSDAG). This is not a new model; it is a subset of the Parameterized CSDF model. Compared with the Parameterized CSDF, the PCSDAG has been simplified and has no cycle. It contains only a single vertex without input edge and the number of firings of this vertex is fixed to 1.

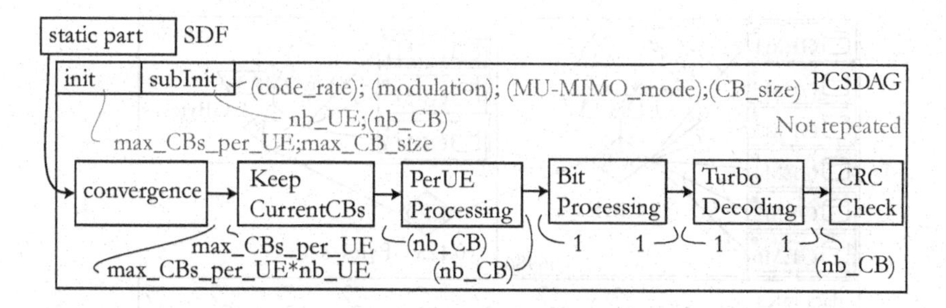

Fig. 8.9 PCSDAG description of the uplink

A PCSDAG description of the LTE PUSCH decoding is shown in Fig. 8.9. Both data token production and consumption of each PCSDAG edge is set by the MAC scheduler. The token production and consumption can either be a single integer value or a pattern of integer values. The PCSDAG has no hierarchy but the parameter (nb_CB) is a template which provides the number of CB for each UE, and all actor parameters are now templates which assign its own parameters: $nb\_CB = \{nb\_CB(UE1), nb\_CB(UE2), nb\_CB(UE3)...\}$ to each CB. For example, a consumption of the number of CBs per UE can contain the pattern $\{10, 5, 3, 1, 1\}$, meaning that the actor will consume 10 data tokens on the first firing, 5 on the second and so on. One millisecond before each subframe, all production and consumption patterns are set by the MAC scheduler. The PCSDAG can then be extended into a single rate DAG (Sect. 3.2.3) and the resulting single rate DAG is scheduled. The DAETs of the LTE PUSCH graph actors also depend on pattern parameters set by the MAC scheduler. The domain of nb_UE is between 0 and 100 UEs (ignoring PUCCH) because each pair of PRBs can be associated with a different UE and the domain of each pattern element in nb_CB is between 1 and 13 CBs. PCSDAG can model LTE PUSCH decoding in a very compact way.

In the next section, the architecture model of the adaptive scheduler is explained.

8.3.3 A Simplified Model of Target Architectures

The processors targeted by this study are the heterogeneous multi-core DSPs introduced in Sect. 5.1.1 and boards interconnecting several of them. They include the 6-core tci6486 and the 3-core tci6488. Combinations of these DSPs (interconnected between test boards with RapidIO serial links for instance) are also targeted architectures. The dataflow graph scheduling techniques naturally handle heterogeneity in data links and operators, treating targets with coprocessors, different DSP frequencies and different communication media. For example, the tci6488 includes a turbo decoding coprocessor that can perform the Forward Error Correction (FEC) [12] decoding part of the PUSCH processing.

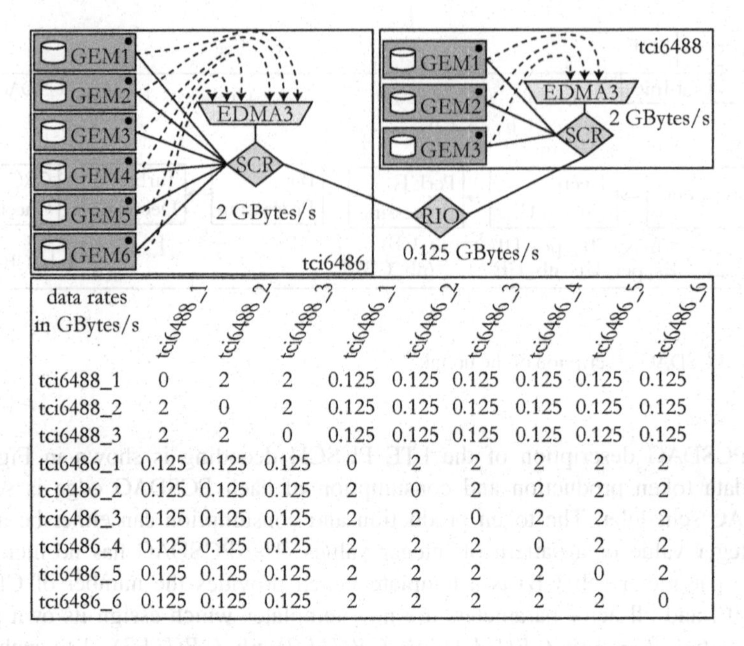

data rates in GBytes/s	tci6488_1	tci6488_2	tci6488_3	tci6486_1	tci6486_2	tci6486_3	tci6486_4	tci6486_5	tci6486_6
tci6488_1	0	2	2	0.125	0.125	0.125	0.125	0.125	0.125
tci6488_2	2	0	2	0.125	0.125	0.125	0.125	0.125	0.125
tci6488_3	2	2	0	0.125	0.125	0.125	0.125	0.125	0.125
tci6486_1	0.125	0.125	0.125	0	2	2	2	2	2
tci6486_2	0.125	0.125	0.125	2	0	2	2	2	2
tci6486_3	0.125	0.125	0.125	2	2	0	2	2	2
tci6486_4	0.125	0.125	0.125	2	2	2	0	2	2
tci6486_5	0.125	0.125	0.125	2	2	2	2	0	2
tci6486_6	0.125	0.125	0.125	2	2	2	2	2	0

Fig. 8.10 Target architecture example in S-LAM and adaptive scheduler matrix model: a tci6488 and a tci6486 connected with a RapidIO serial link

Using S-LAM (Chap. 5) in the adaptive scheduler would not be realistic in terms of memory and computing cost. Therefore, a very simple architecture model is used, In this model, the DAET of actors in the single rate DAG are linked to operator types, enabling operator heterogeneity. A matrix is created where each oriented pair of source and sink operators is associated with a transfer speed in Gbytes/s. This matrix is displayed in Fig. 8.10 and its results are compared to that of S-LAM for an example consisting of one tci6486 and one tci6488.

8.3.4 Adaptive Multi-Core Scheduling of the LTE PUSCH

An obvious solution to efficiently schedule a dynamic algorithm onto a heterogeneous architecture is to schedule all possible configurations at compile-time and then switch between the pre-computed schedules at run-time. The limits of such an approach are now analyzed.

8.3.4.1 The Limits of Pre-Computed Scheduling

Assigning a given number of CBs to UEs is a similar problem to partitioning the number of CBs into a sum of integers. The problem of integer partitioning is illustrated by Ferrer diagrams in Fig. 8.11. Given a number N_{CB} of CBs to assign,

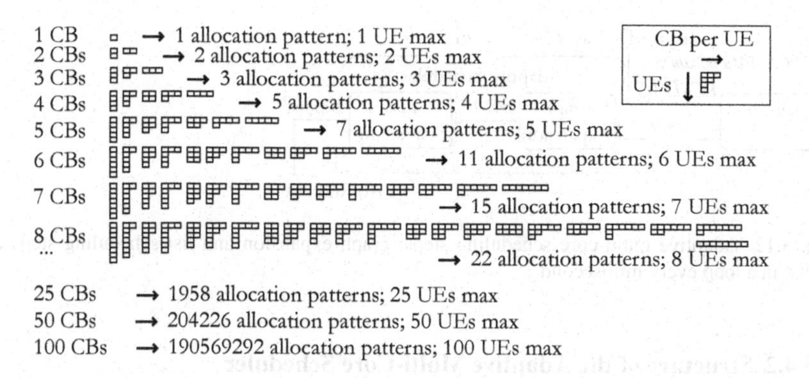

Fig. 8.11 The problem of allocating CBs to UEs is equivalent to integer partitions represented here in Ferrer diagrams

a Ferrer diagram gives the possible number of different allocation configurations $p(N_{CB})$, i.e. the possible number of different graphs to schedule and map for a given quantity of CBs. The number of CBs N_{CB} allocated every millisecond also varies between 0 and $N_{CB_{MAX}}$. In Sect. 2.3.4, the maximum number of pairs of PRBs per subframe was shown to be a constant for each eNodeB and is fixed between 6 and 100 depending on the LTE system bandwidth. $N_{CB_{MAX}}$ is equal to the maximum number of pairs of PRBs per subframe because PRB pair is the minimum resource released to a code block. The total number of different possible graphs for a given eNodeB PUSCH is:

$$P(N_{CB_{MAX}}) = \sum_{i=1}^{N_{CB_{MAX}}} p(i). \tag{8.1}$$

For the simplest case of a 1.4 MHz LTE system bandwidth with only a maximum of 6 CBs to allocate, the number of possible graphs is 29. It is feasible to pre-compute and store the scheduling of these 29 graphs. However, the number of graphs increases exponentially with N_{CB}; the still relatively simple case of 3 MHz with a maximum of 12 CBs generates 271 graphs. Considering that for $P(50) = 1295970$, it can be concluded that the scheduling of all LTE PUSCH cases with bandwidth higher than 3 MHz cannot realistically be statically scheduled.

It must be noted that the above calculation of cases for each bandwidth only takes into account topology modifications and not execution time modifications due to the variable size of the CBs. The next section explains an adaptive scheduler that can efficiently schedule PUSCH despite topology and CB size variations.

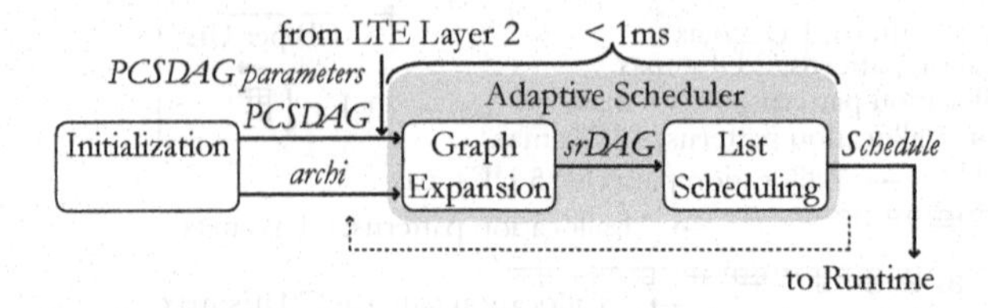

Fig. 8.12 Adaptive multi-core scheduling steps: graph expansion and list scheduling steps are called in a loop every millisecond

8.3.4.2 Structure of the Adaptive Multi-Core Scheduler

The adaptive multi-core scheduler algorithm is illustrated in Fig. 8.12. There is an initialization phase that generates the PCSDAG object from a manual description and pre-computes the graph parameter expressions. The following processing steps are called each millisecond when the MAC scheduler changes the CB allocation. These scheduler steps consist of 2 operations:

- The `graph expansion` which transforms the PCSDAG into a single rate DAG with shape dependent on the PCSDAG parameters.
- The `list scheduling algorithm` which maps each actor in the single rate DAG to an operator (core or coprocessor) in the architecture.

The goal of the initialization step of the adaptive scheduler is to maximally reduce the loop step complexity. The expressions of the PCSDAG edge productions and consumptions are then parsed and converted into a Reverse Polish Notation (RPN) stack using the Shunting-yard algorithm [13]. RPN specifies the order of expression evaluation in an efficient bracket-less expression. This technique allows the adaptive scheduler to manipulate efficiently the PCSDAG patterns, which is essential in a run-time scheduler.

The output of the adaptive scheduler feeds a run-time procedure. This procedure may implement a multi-core API (OpenCL [14], Multi-core Association [15]). The sum of graph expansion and list scheduling execution times must stay under 1 ms for the scheduler to be an approach of interest in LTE PUSCH processing. These two operations are detailed in the following Sections.

8.3.4.3 Adaptive Expansion of PCSDAG into Single Rate DAG

A single rate DAG (Sect. 3.1.2) is a graph with no cycle, in which each edge has equal data production and consumption. All single rate DAG actors are instances of PCSDAG actors. Each single rate DAG actor is fired only once.

Figure 8.13 gives four examples of single rate DAG graphs generated from the expansion of the PCSDAG in Fig. 8.9 with different allocation schemes. The number of (possible) vertices in a single rate DAG may be calculated by:

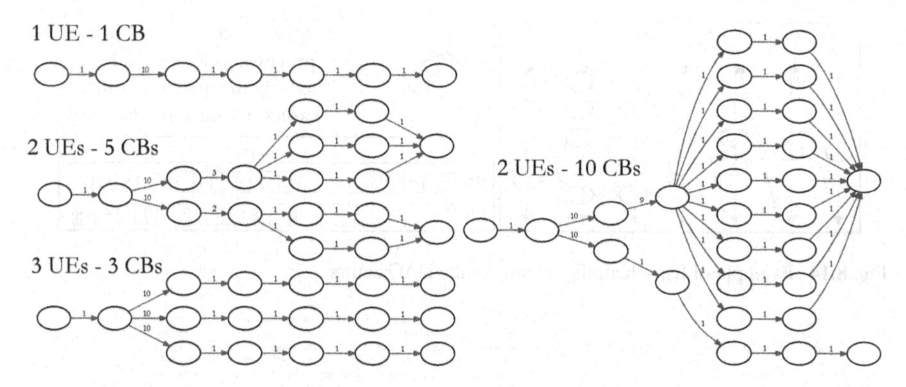

Fig. 8.13 Examples of single rate DAG graphs generated from the PCSDAG description in Fig. 8.9

$$V = 2 + 3 * nb\_UE + 2 * N_{CB}. \tag{8.2}$$

Thus, for the PUSCH example, the maximal size of the corresponding single rate DAG is 502 vertices with 100 UEs and 100 CBs allocated in the subframe. To generate the second case shown in the figure (2 UEs–5 CBs), we have $nb\_UE = 2$, $CBs\_UE = \{2, 3\}$, and $max\_CBs\_UE = 3$ from the PCSDAG graph (of Fig. 8.9).

The adaptive scheduler algorithm calculates the number of firings of each actor, using the productions and consumptions of the incoming edges (as computed from RPN expression stacks) and the firings of the preceding actors only. This calculation is then stored in the Basis Repetition Vector (BRV) presented in Sect. 3.2. The absence of a loop in the graph and the existence of only one source vector with a single firing enables this "time-greedy" method in $O(|E| + |V|)$, where $|E|$ and $|V|$ are the number of edges and the number of vertices in the single rate DAG respectively. There is no schedulability checking in the expansion step. Once the BRV has been calculated and the single rate DAG vertices have been created, the appropriate edges are added to interconnect them.

This technique may be contrasted with SDF. SDF schedulability checking and expansion into single rate SDF requires a study of the null space of the topology matrix presented in Sect. 3.2. The method, described in Algorithm 3.1, has a time complexity of $O(|V|^2|E|)$. Although yielding good results, the time cost of this operation excludes its execution in a realistic run-time scheduler.

After the expansion of the PCSDAG graph, the resulting single rate DAG is then ready for the list scheduling step. This operation is explained in the next Section.

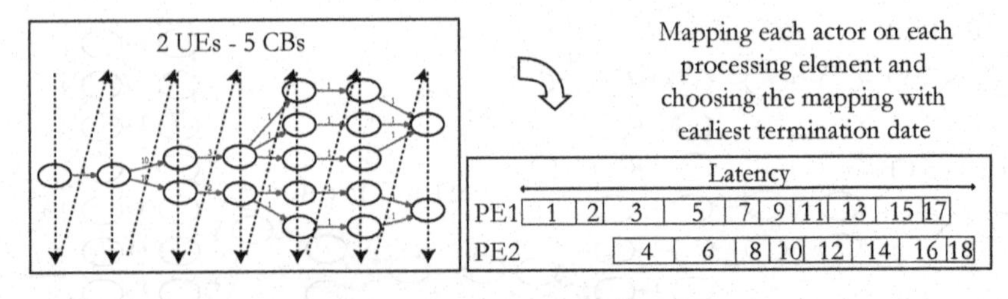

Fig. 8.14 Example of list scheduling of single rate DAG actors

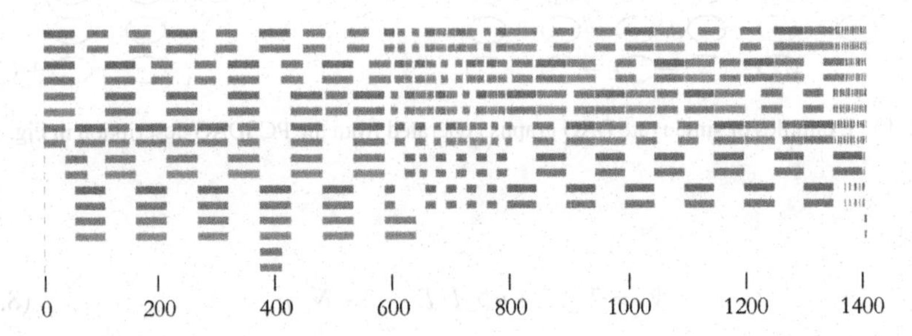

Fig. 8.15 Example of a schedule Gantt chart: case with 100 CBs and 100 UEs on the architecture from Fig. 8.10 with one core reserved for the scheduling

8.3.4.4 List Scheduling of the Single Rate Directed Acyclic Graph

The static list scheduling operation used is a simplified version of the "time-greedy" algorithm described in Sect. 4.4.3. The list scheduling process is illustrated in Fig. 8.14 and an example of a Gantt chart generated by the adaptive scheduler (Fig. 8.15) gives an idea of the resulting scheduling complexity.

As may be seen in Fig. 8.14, there is no actor reordering process before the list scheduling algorithm is executed. In the LTE PUSCH case, the consequences for the scheduling quality are limited because all the paths from the first actor to actors without successors are equivalent, except for the DAETs of the actors. A suboptimal input list results in a multiplication of the latency by a factor $\lambda \leq 2 - 1/n$ where n is the number of target cores [16]. This result is an approximation as the calculation ignores the data transmission latencies. The single rate DAG is not reordered in the adaptive scheduler is because the topological order of the single rate DAG with its absence of cycle, is naturally deduced from the PCSDAG. Searching for another topological order for the single rate DAG would require the prohibitively time consuming actor list construction with $O(|V| \cdot log(|V|))$ time complexity presented in Sect. 4.4.3. Without this reordering operation, the adaptive scheduler cost is perfectly linear with the input graph size and architecture size. The next Section illustrates this behavior for an implementation of the adaptive scheduler.

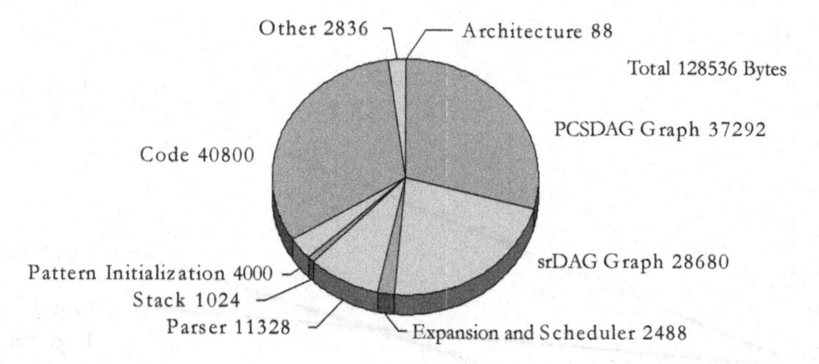

Fig. 8.16 Memory footprint, in Bytes, of the statically allocated adaptive scheduler

8.3.5 Implementation and Experimental Results

The adaptive scheduler implementation genericity, compactness and speed were developed with special care, due to their vital role. The entire code, including private members and inlined accessors, is written in C++.

8.3.5.1 Memory Footprint of the Adaptive Scheduler

In order to reduce execution time, the adaptive scheduler contains no dynamic allocation. Its memory footprint (Fig. 8.16) is only 126 kBytes. Each c64x+ core of the tci6486 and the tci6488 processors of Sect. 5.1.1 has an internal local L2 memory of 608 kBytes and 500–1,500 kBytes respectively. Thus, the memory footprint is sufficiently small to fit within the internal local L2 memory of a single core of either processor.

Half of the memory footprint is used by the graphs. The relatively large size of the PCSDAG in memory is due to the patterns stored in RPN. The single rate DAG graph has a size sufficient to contain any possible LTE PUSCH configuration. One third of the footprint is used by the code.

8.3.5.2 Impact of Graph and Architecture Sizes on the Adaptive Scheduler Speed

Figure 8.17 shows that the execution time of the Adaptive Scheduler increases linearly with the graph size. The graph expansion time for very small single rate DAGs is due to the constant RPN parameter evaluation. The worst case execution time is less than 950,000 cycles, enabling real-time execution on a c64x+ at 1 GHz. However, a load of 95 % for the scheduling core is unwise for a real application. For this case, further optimizations would be necessary to lower the load.

In Fig. 8.18, adaptive scheduler execution time is shown to increase linearly with the number of cores in the architecture. The graph shows the case where the single

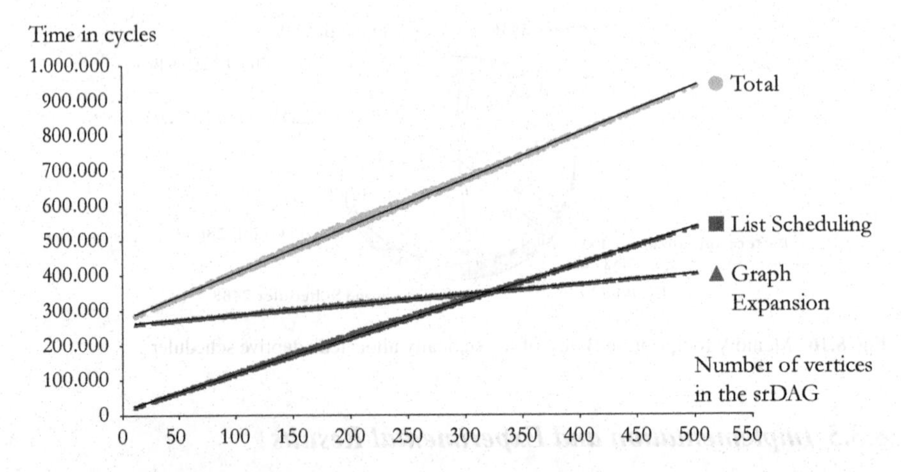

Fig. 8.17 Impact of single rate DAG graph size on the LTE PUSCH scheduling execution time on the architecture of Fig. 8.10 with one core reserved for the scheduling

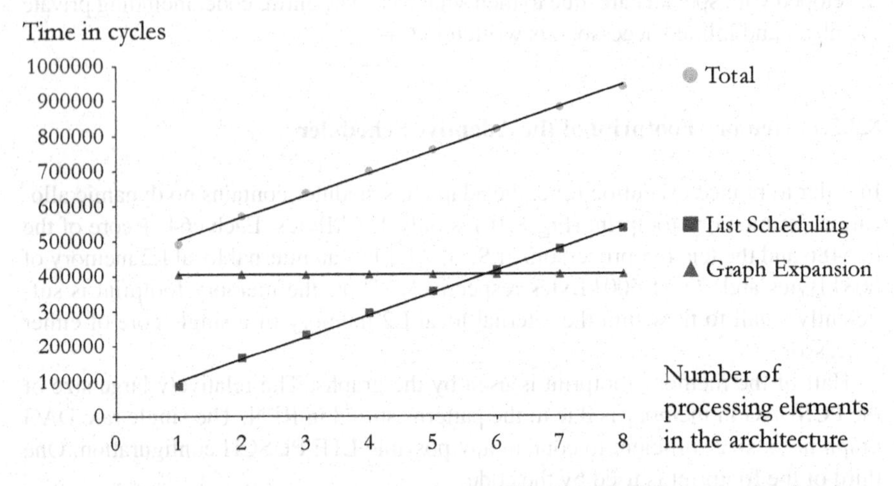

Fig. 8.18 Impact of the number of operators on the LTE uplink scheduling execution time in an LTE PUSCH worst case

rate DAG size is fixed to 502 and the number of cores varies. Moreover, it may be seen from this figure that the graph expansion time only varies with PCSDAG parameters, and does not depend on architecture size. The maximum number of cores targeted by the present implementation with a DSP at 1 GHz is 9 (this includes a dedicated core for scheduling). Specific optimizations for Texas Instruments chips with intrinsic and pragma can improve this result.

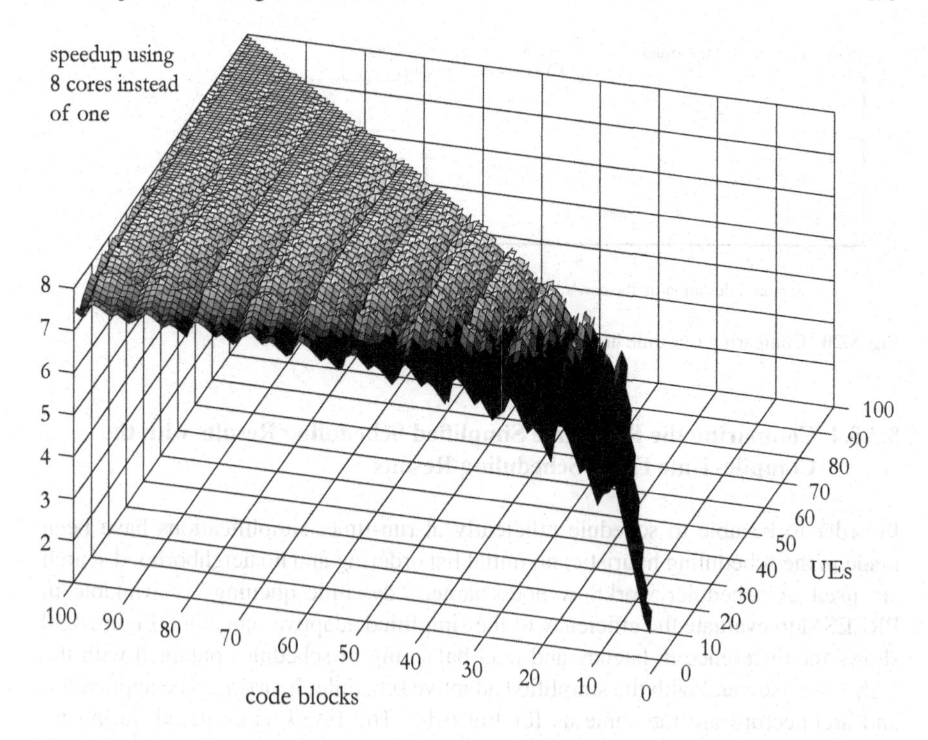

Fig. 8.19 Schedule speedup versus number of CBs and UEs using architecture from Fig. 8.10

8.3.5.3 Evaluating the Limits of the Target Architecture

The latency of the dynamic part of LTE PUSCH depends greatly on the instantaneous number of decoded CBs, the size of those CBs that influences their decoding time, and the instantaneous number of communicating UEs. Figure 8.19 shows the increase in speed obtained with the use of multiple cores, instead of a single 700 MHz c64x+. For this test, each CB has a size of one pair of PRBs and can transport 760 bits. The speed increase approaches the theoretical maximum of $(6*0.7 + 2*1)/0.7 = 8.8$ times when the single rate DAG becomes larger and exercises more data parallelism. The architecture of Fig. 8.10 has one core of the tci6488 dedicated to scheduling and no coprocessor, and is likely to be sufficient to decode the LTE PUSCH dynamic part in 1 ms for 50 CBs and 50 UEs, i.e. for the 10 MHz bandwidth case. The execution time of the static part of the algorithm is not taken into account in this simulation. The problem of the multi-core partitioning of the static part then becomes easier as it can be solved at compile-time. However, this operation will need to be pipelined with the dynamic part in order to respect the 1 ms execution time limit. The load of the core executing the scheduler will need to be limited as discussed above and the dynamic part combined with the static part.

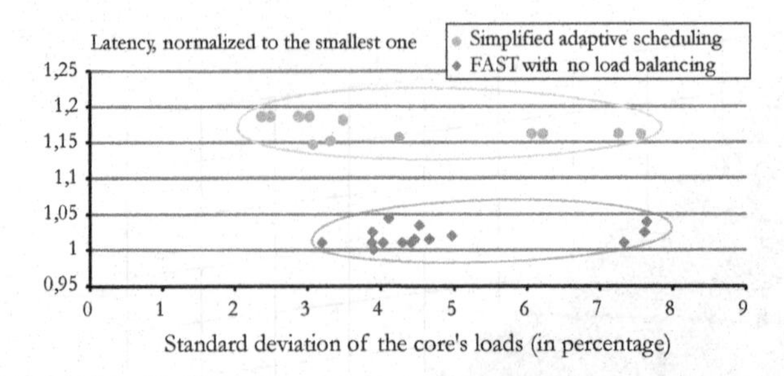

Fig. 8.20 Comparing run-time heuristic schedules with FAST schedules

8.3.5.4 Comparing the Run-Time Simplified Scheduling Results with the Compile-Time FAST Scheduling Results

In order to be able to schedule efficiently at run-time, simplifications have been made in the scheduling heuristic: no initial list ordering and no neighborhood search are used. A scheduler workflow node named "dynamic queuing" is available in PREESM to evaluate the efficiency of the simplified adaptive scheduler. Figure 8.20 shows the difference in latency and load balancing of schedules obtained with the FAST heuristic and with the simplified adaptive scheduler heuristic. The application and architecture are the same as for Fig. 6.11. The FAST is executed during 60 seconds without balancing the loads (Sect. 6.4.3). It may be seen in Fig. 8.20 that using simplified adaptive scheduler heuristic instead of FAST, the latency is increased by about 15 % while the load balancing remains approximately stable (it is even improved in some cases). Given the large difference in processing time, the simplified adaptive scheduler heuristic is thus performing well.

In the next section, the PDSCH model is studied and shown to be close to the PUSCH model.

8.4 PDSCH Model for Adaptive Scheduling

In the rapid prototyping model of downlink decoding displayed in Fig. 7.6, the channel encoding shows many similarities with the uplink channel decoding shown in Fig. 7.8. However, the significant difference is that the number of TBs per UE varies between one and two and must be chosen, and this parameter must be resolved adaptively for each UE.

PDSCH decoding dominates the LTE downlink decoding. A PDSCH PCSDAG model for adaptive scheduling is proposed in Fig. 8.21. This model is topologically close to that of PUSCH, shown in Fig. 8.9. Actors from Fig. 7.6 have been clustered to reduce their number. TB bit processing gathers CRC additions and code block segmentations; turbo encoding embeds CB CRC additions; and channel coding contains concatenation, scrambling and constellation mapping. nb_TB gives the total

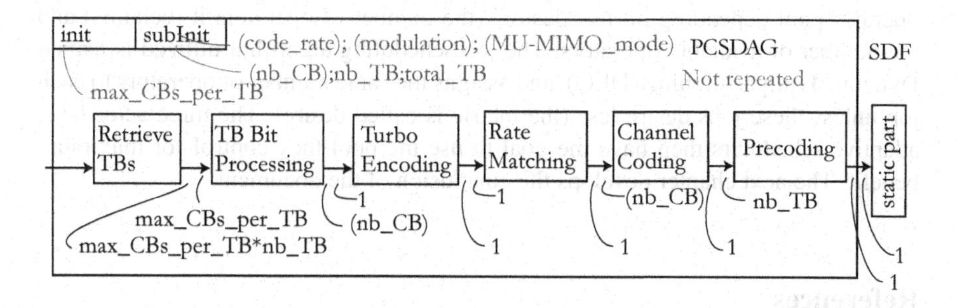

Fig. 8.21 PCSDAG model of the PDSCH LTE channel

number of TBs to encode in the subframe and (nb_CB) the pattern of the number of CBs in each TB.

The maximum graph size is 803 actors. Scheduling this graph adaptively is thus more costly than scheduling PUSCH and further optimizations of the scheduler will be necessary to minimize this cost. The next section briefly addresses the combination of uplink decoding, downlink encoding and RACH-PD for entire LTE physical layer execution of an eNodeB.

8.5 Combination of Three Actor-Level LTE Dataflow Graphs

Currently, the majority of signal processing applications have similar structures with three distinct levels of behavior. Reusing the terminology from Sect. 4.1.1, these levels are:

- Instruction-level with both control and data streams, efficiently described by imperative MoCs and efficiently executed on Von Neumann machines from these descriptions.
- Data-dominated actor-level, which is well suited for distributed architectures, provided sufficient data rate is available between operators. Models at this level are called concurrency models in the literature.
- Loosely-coupled task-level which is dominated by control streams and is well suited for modeling with state machines.

The LTE eNodeB physical layer follows this model structure with three actor-level tasks: RACH-PD, uplink and downlink which are connected by the control-dominated MAC layer, and contain actors which locally mix control and data dependencies. The appropriate global execution scheme for the LTE physical layer is certainly a two-level hierarchical scheduler (Sect. 8.1.1) with a master/slave task-level scheduler and three actor-level adaptive schedulers. The technique for task-level scheduling could be based on [2] where a given task is associated with a particular operator pool depending on its "desire" (the number of operators it reclaims) and the number of available operators. The job scheduling algorithm utilized is named Dynamic Equipartitioning (DEQ), and assigns the same number of operators to each job unless these jobs desire less (the metric is called desire). The three actor-level adaptive schedulers then have the goal to use the pool they control for maximum benefit. The next chapter develops the conclusion of the document.

References

1. Bell D, Wood G (2009) Multicore programming guide. Technical report, Texas Instruments
2. He Y, jing Hsu W, Leiserson CE (2006) Provably efficient two-level adaptive scheduling. http://citeseerx.ist.psu.edu/viewdoc/summary? doi:10.1.1.94.7729
3. Ha S, Lee EA (1997) Compile-time scheduling of dynamic constructs in dataflow program graphs. IEEE Trans Comput 46:768–778
4. Grandpierre T, Sorel Y (2003) From algorithm and architecture specifications to automatic generation of distributed real-time executives: a seamless flow of graphs transformations. In: MEMOCODE '03, pp. 123–132, 2003
5. Piat J, Bhattacharyya SS, Pelcat M, Raulet M (2009) Multi-core code generation from interface based hierarchy. DASIP 2009
6. TMS320 DSP/BIOS Users Guide (SPRU423F)
7. Pelcat M, Aridhi S, Nezan JF (2008) Optimization of automatically generated multi-core code for the LTE RACH-PD algorithm. 0811.0582 (2008), DASIP 2008, Bruxelles : Belgique.http://arxiv.org/abs/0811.0582
8. RapidIO. http://www.rapidio.org/home/
9. Sriram S, Bhattacharyya SS (2009) Embedded multiprocessors: scheduling and synchronization, 2nd edn. CRC press, Boca Raton
10. Moreau S, Aridhi S, Raulet M, Nezan JF (2007) On modeling the RapidIO communication link using the AAA methodology. In: DASIP, 2007
11. Pelcat M, Piat J, Wipliez M, Nezan JF, Aridhi S (2009) An open framework for rapid prototyping of signal processing applications. EURASIP J Embed Syst (2009)
12. Sesia S, Toufik I, Baker M (2009) LTE, The UMTS long term evolution: from theory to practice. Wiley, New York
13. Dijkstra E (1960) Algol 60 translation. Supplement, Algol 60 Bulletin **10** (1960)
14. OpenCL. http://www.khronos.org/opencl/
15. The Multicore Association. http://www.multicore-association.org/home.php
16. Graham RL (1966) Bounds for certain multiprocessing anomalies. Bell Syst Tech J 45(9):1563–1581

Chapter 9
Conclusion

The most recent algorithms and architectures for embedded systems have become so complex that a `system-level` view of projects from the early design stages to the implementation is now necessary in order to avoid bad design choices and to meet deadlines. A multi-core DSP implementation of a 3GPP LTE base station is representative of these new complex systems which require high optimization. The software of such a heterogeneous embedded distributed system cannot be efficiently developed without a special development chain based on rapid prototyping and system-level simulation. In this book, `software rapid prototyping` methods were introduced to replace certain tedious and sub-optimal steps of the present test-and-refine methodologies for embedded software development. These techniques were applied to the study of a 3GPP LTE base station physical layer.

Building a high-level view of a distributed and heterogeneous embedded system brings `new challenges` compared with sequential software development chains. An intimidating number of complex problems arise when algorithm actors are assigned to architecture operators and algorithm data transfers are assigned to architecture communication nodes. A compromise must be made between model expressivity and power of analysis. Thankfully, a large amount of relevant literature exists, due to the search for heuristics to solve similar parallelization problems such as organizing factories, and projects. In this book, heuristics from the literature are presented and enhanced for integration in automatic prototyping processes for heterogeneous embedded systems.

A system-level view requires `new algorithm and architecture models`. When extracting algorithm parallelism from sequential imperative code, it is tempting to make the transition between sequential and distributed machines seamless. However, imperative code introduces useless dependencies to an algorithm and adds a new and unnecessary parallelism extraction problem to the existing challenges. In this document, several dataflow algorithm models are presented and their use to model LTE is justified and explained. They naturally express algorithm parallelism and have already been intensely studied. A new simple and expressive System-Level Architecture Model is also introduced, enabling high speed simulations.

M. Pelcat et al., *Physical Layer Multi-Core Prototyping*,
Lecture Notes in Electrical Engineering 171, DOI: 10.1007/978-1-4471-4210-2_9,

Programming a distributed embedded system at high-level, designers have several constraints to respects and costs to optimize. Moreover, these costs are highly dependent on the system. An embedded software development chain must adapt to these different costs and assess the quality of the resulting solutions: it must consist of a framework embedding functionalities rather than a monolithic tool. In this document, the PREESM rapid prototyping framework was presented. It embeds many functionalities, each with multiple parameters. Importantly, a graphical schedule quality assessment chart was also introduced. It displays the parallelism available in an algorithm in terms of latency and shows the present use of this parallelism in the system.

Finally, code generation from model-based descriptions was explained. LTE physical layer algorithms were detailed and previously evoked methods were applied to the implementation of LTE on multi-core DSPs. These approaches revealed limitations in the compile-time methods for multi-core scheduling. Depending on the particular algorithm used, the compile-time or the run-time multicore scheduling method should be chosen. The resulting run-time system can be based on different multi-core scheduling strategies and different execution schemes and must balance the correct use of the parallel architecture with the scheduling overhead.

Glossary

3GPP Third Generation Partnership Project

3G Third Generation Telecommunication System

AADL Architecture Analysis and Design Language

AAM Algorithm Architecture Matching

ACK HARQ Acknowledgement

AIF Antenna Interface

ALP Actor Level Parallelism

ALU Arithmetic and Logic Unit

APEG Acyclic Precedence Expansion Graph

APN Arbitrary Processor Network

ARQ Automatic Repeat reQuest

ASAP As Soon As Possible

AWGN Additive White Gaussian Noise

BDF Boolean Dataflow Graph

BNP Bounded Number of Processors

BPSK Binary Phase Shift Keying

BRV Basis Repetition Vector

BSR Buffer Status Report

C-RNTI Random Access Radio Network Temporary Identifier

CAZAC Constant Amplitude Zero AutoCorrelation

M. Pelcat et al., *Physical Layer Multi-Core Prototyping*,
Lecture Notes in Electrical Engineering 171, DOI: 10.1007/978-1-4471-4210-2,

CB Code Block

CDD Cyclic Delay Diversity

CDM Code Division Multiplexing

CIR Channel Impulse Response

CISC Complex Instruction Set Computer

CN Communication Node

CPN Critical Path Node

CP Critical Path

CP Cyclic Prefix

CQI Channel Quality Indicator

CRC Cyclic Redundancy Check

CSDF Cyclo Static Dataflow Graph

DAET Deterministic Actor Execution Time

DAG Directed Acyclic Graph

DCI Downlink Control Information

DDF Dynamic Dataflow Graph

DDR Double Data Rate

DE Discrete Event

DFT Discrete Fourier Transform

DM RS Demodulation Reference Signal

DMA Direct Memory Access

DSL Digital Subscriber Line

DSP Digital Signal Processor

E-UTRAN Evolved Universal Terrestrial Radio Access Network

EMAC Ethernet Media Access Controller

eNodeB evolved NodeB

EPC Evolved Packet Core

ETSI European Telecommunications Standards Institute

FAST Fast Assignment and Scheduling of Tasks

FDD Frequency Division Duplex

FEC Forward Error Correction

FFT Fast Fourier Transform

FIFO First-In First-Out

FSM Finite State Machine

GPU Graphical Processing Unit

GSM Global System for Mobile Communications

GT Guard Time

HARQ Hybrid Automatic Repeat reQuest

HD-FDD Half-Duplex FDD

HDL Hardware Description Language

HSPA High Speed Packet Access

HSS Home Subscriber Server

IBN In-Branch Node

IBSDF Interface-Based Hierarchical Synchronous Dataflow Graph

ICI Inter Carrier Interference

IDF Integer Dataflow Graph

IDL Interface Description Languages

IFFT Inverse Fast Fourier Transform

ILP Instruction-Level Parallelism

IMT International Mobile Telecommunications

IP Intellectual Property

IP Internet Protocol

ISA Instruction Set Architecture

ISI Inter Symbol Interference

ITRS International Technology Roadmap for Semiconductors

ITU-R International Telecommunication Union, Radio Communication Sector

KPN Kahn Process Network

LTE Long Term Evolution

M-sequence Maximum length sequence

MAC Medium Access Control

MCS Modulation and Coding Scheme

MIB Master Information Block

MIMD Multiple Instruction Multiple Data

MIMO Multiple Input Multiple Output

MISD Multiple Instruction Single Data

MLD Maximum Likelihood

MME Mobility Management Entity

MMSE Minimum Mean-Square Error

MoC Model of Computation

MPSoC Multi-Processor System-on-Chip

MRC Maximal-Ratio Combining

MU-MIMO Multi-User MIMO

NACK HARQ Non Acknowledgement

NAS Non Access Stratum

NLOS Non-line-of-sight

NORMA NO Remote Memory Access

NUMA Non Uniform Memory Access

OBN Out-Branch Node

OFDMA Orthogonal Frequency Division Multiplexing Access

OOP Object-Oriented Programming

OSI Open Systems Interconnection

P-GW Packet Data Network Gateway

PAPR Peak to Average Power Ratio

PBCH Physical Broadcast Channel

PCFICH Physical Control Format Indicator Channel

PCRF Policy Control and charging Rules Function

PCSDAG Parameterized Cyclo Static Directed Acyclic Graph

PDCCH Physical Downlink Control Channel

PDCP Packet Data Convergence Protocol

PDSCH Physical Downlink Shared Channel

PDU Protocol Data Unit

PFAST Parallel Fast Assignment and Scheduling of Tasks

PHICH Physical Hybrid ARQ Indicator Channel

PMCH Physical Multicast Channel

PMI Precoding Matrix Indicator

PN Process Network

PRACH Physical Random Access Channel

PRB Physical Resource Block

PREESM Parallel and Real-time Embedded Executives Scheduling Method

PSDF Parameterized Synchronous Dataflow Graph

PSS Primary Synchronization Signal

PUCCH Physical Uplink Control Channel

PUSCH Physical Uplink Shared Channel

QoS Quality of Service

RA-RNTI Random Access Radio Network Temporary Identifier

RACH-PD Random Access Channel Preamble Detection

RAM Random Access Memory

RAR Random Access Response

RE Resource Element

RF Radio Frequency

RISC Reduced Instruction Set Computer

RI Rank Indicator

RLC Radio Link Control

RPN Reverse Polish Notation

RS Reference Signal

RTOS Real-Time Operating System

S-GW Serving Gateway

S-LAM System-Level Architecture Model

SAE Society of Automotive Engineers

SAE System Architecture Evolution

SC-FDMA Single Carrier-Frequency Division Multiplexing Access

SCR Switched Central Resource

SDF4J Synchronous Dataflow for Java

SDF Synchronous Dataflow Graph

SDMA Spatial Division Multiple Access

SDU Service Data Unit

SD Sphere Decoder

SFBC Space-Frequency Block Coding

SIB System Information Block

SIMD Single Instruction Multiple Data

SINR Signal-to-Interference plus Noise Ratio

SISD Single Instruction Single Data

SNR Signal-to-Noise Ratio

SRS Sounding Reference Signal

SR Scheduling Request

SSE Streaming SIMD Extensions

SSS Secondary Synchronization Signal

STBC Space-Time Block Coding

TB Transport Block

TDD Time Division Duplex

ThLP Thread-Level Parallelism

TLM Transaction Level Modeling

TLP Task-Level Parallelism

TTI Transmission Time Interval

UE User Equipment

UMA Uniform Memory Access

UMTS Universal Mobile Telecommunications System

UNC Unbounded Number of Clusters

VLIW Very Long Instruction Word

VoIP Voice over IP

WCET Worst-Case Execution Time

XSLT Extensible Stylesheet Language Transformation

ZC Zadoff-Chu Sequence

ZF Zero Forcing

Index

M. Pelcat et al., *Physical Layer Multi-Core Prototyping*,
Lecture Notes in Electrical Engineering 171, DOI: 10.1007/978-1-4471-4210-2,